NUMBERS RULE

NUMBERS RULE

The Vexing Mathematics of Democracy,
from Plato to the Present

George G. Szpiro

PRINCETON UNIVERSITY PRESS

PRINCETON AND OXFORD

Published by Princeton University Press, 41 William Street, Princeton, New Jersey 08540
In the United Kingdom: Princeton University Press, 6 Oxford Street, Woodstock,
Oxfordshire OX20 1TW

Library of Congress Cataloging-in-Publication Data

Szpiro, George, 1950–
Numbers rule : the vexing mathematics of democracy, from Plato to the
present George G. Szpiro / George G. Szpiro.
p. cm.
Includes bibliographical references and index.
ISBN 978-0-691-13994-4 (alk. paper)
1. Mathematics—Anecdotes. I. Title.
QA99.S97 2010
510—dc22   2009028615

British Library Cataloging-in-Publication Data is available

This book has been composed in Century
Printed on acid-free paper. ∞
press.princeton.edu
Printed in the United States of America
1   3   5   7   9   10   8   6   4   2

Dedicated to Sarit and Nir
At the start of the new chapter in their lives.
(Until 120!)

CONTENTS

PREFACE

It may come as a surprise to many readers that our democratic institutions and the instruments to implement the will of the people are by no means foolproof. In fact, they may have strange consequences. One example is the so-called Condorcet Paradox. Named after the eighteenth-century French nobleman Jean-Marie Marquis de Condorcet, it refers to the surprising fact that majority voting, dear to us since times immemorial, can lead to seemingly paradoxical behavior. I do not want to let the cat out of the bag just yet by giving away what this paradox is. Suffice it to say for now that this conundrum has kept mathematicians, statisticians, political scientists, and economists busy for two centuries—to no avail. Worse, toward the middle of the twentieth century, the Nobel Prize winner Kenneth Arrow proved mathematically that paradoxes are unavoidable and that *every* voting mechanism, except one, has inconsistencies. As if that were not enough, a few years later, Allan Gibbard and Mark Satterthwaite showed that every voting mechanism, except one, can be manipulated. Unfortunately, the only method of government that avoids paradoxes, inconsistencies, and manipulations is a dictatorship.

There is more bad news. The allocation of seats to a parliament, say to the U.S. Congress, poses further enigmas. Since delegations must consist of whole persons, they must be integer numbers. How many representatives should be sent to Congress, for example, if a state is due 33.6 seats? Should it be thirty-three or thirty-four congresspeople? Simple rounding will usually not work because in the end, the total number may not add exactly to the required 435 congressmen. Alternative suggestions have been made, in the United States as well as in other countries, but they are fraught with problems, some methods favoring small states others favoring big states. And that is not the worst of it. Under certain circumstances, some states may actually *lose* seats if the size of the House is *increased*. (This bizarre situation has become notorious under the designation Alabama Paradox.) Other absurdities are known as the Population Paradox and the New State Paradox. Politicians, scientists, and the courts have been battling with the problems for centuries. But similarly to Arrow's Theorem, it finally turned out that no solution to the problem exists. The mathematicians Peyton Young and Michel Balinski proved that there are no *good* or *correct* methods to allocate seats to Congress or any other parliament.

This book is an elucidation and a historical account of the problems and dangers that are inherent in the most cherished instruments of democracies. The narrative starts two and a half millennia ago, with ancient the Ancient Greek and Roman thinkers Plato and Pliny the Younger, continues to the churchmen of the Middle Ages Ramon Llull and Nicolaus Kues, goes on to heroes and victims of the French Revolution Jean-Charles de Borda and the Marquis de Condorcet, and from there turns to the Founding Fathers and ends with modern-day scholars like Arrow, Gibbard, Satterthwaite, Young, and Balinski

I have written this book for a general readership, my aim being to introduce readers to the subject matter in an entertaining way. Hence it is by no means a textbook but may serve as accompanying literature for a more rigorous course in political science, economics, administration, philosophy, or decision theory. Much weight has been given to the personalities who have been involved during the past two and a half millennia in the endeavor to understand the problems and in the attempts to correct them. In order not to disturb the flow of the text, however, material about the *dramatis personae* and their times is often consigned to an additional reading section at the end of the chapters.

While expounding on the blithe aspects of the subject matter, I do not at all make light of its importance and of the serious difficulties. But even though the issues are very deep, the arguments never involve anything but the most basic mathematics. In fact, readers need not have more knowledge of mathematics than what they learned in junior high school. It is inherent in the problems that the mathematics never moves beyond simple arithmetic. But let the reader not be deceived by this apparent simplicity; the questions are surprisingly deep and the arguments amazingly sophisticated.

Many people have been very generous with help and guidance. Unfortunately, I cannot list them all here for a simple, if unpleasant, reason: my e-mail archive got corrupted and I have been unable to reestablish the correspondence with people whose advice I sought while writing this book. I sincerely apologize. Those whose e-mails I could find are: Kenneth Arrow, Michel Balinski, Daniel Barbiero, Anthony Bonner, Robert Inman, Eli Passow, Friedrich Pukelsheim, Christoph Riedweg, and Peyton Young. I would also like to thank Vickie Kearn, Anna Pierrehumbert, and Heath Renfroe from Princeton University Press and freelancer Dawn Hall for diligent editing; three referees for painstaking reviews; and, as always, my agent Ed Knappman who also came up with the book's catchy subtitle.

While revising the manuscript, I spent a one-month "sabbatical" at the Rockefeller Foundation's Bellagio Study and Conference Center. Bellagio is a beautiful village on the shore of Lake Como in Italy that was once the property of none other than the hero of chapter 2, Pliny the Younger. He described it as such: "Set high on a cliff . . . it enjoys a broad view of the lake which is divided in two by the ridge on which it stands. . . . From the spacious terrace, the descent to the lake is gentle. . . ." You may gather from this that Pliny was in love with the place, and so were my wife and I. I am very grateful to the Rockefeller Foundation for having afforded us the occasion to walk in the footsteps of Pliny the Younger . . . and to put the finishing touches to this book.

Jerusalem, May 2009

NUMBERS RULE

CHAPTER ONE
THE ANTI-DEMOCRAT

Plato, son of Ariston and Perictione, has been called the greatest of Greek philosophers by his admirers and chastised as the worst anti-democrat by his detractors. Socrates' most brilliant student, Plato devoted his life to studying and teaching, to exploring the meaning of life, inquiring into the nature of justice, and pondering how to be a better person.

His real name may have been Aristocles with the nickname Plato—meaning broad—given to him because of his wide forehead or because of his wide-ranging intellectual pursuits. He was born in 427 BC in or near Athens. Plato had two brothers, Glucon and Adeimantus, and a sister Potone. When he was still a boy, his father died and his mother married her uncle Pyrilampes, with whom she had another son, Plato's half-brother Antiphon. Plato received a first-class education in gymnastics, music, poetry, rhetoric, and mathematics, and tried his hand as a playwright. When he became older and a bit more knowledgeable about poetry, he would burn all his plays.

Like many of his peers, the intellectually curious young man was drawn to the circle of students around the philosopher Socrates. It was the best show in town, certainly more interesting than the tedious sessions of the assembly, the council, or the courts. Many sons of aristocratic Athenian families flocked to the philosopher who taught them about the proper manner of reasoning. Socrates considered himself not so much a teacher spouting opinions and truths but a midwife, like his mother had been, helping his pupils bring forth knowledge that was present in their mind but hidden from their consciousness. The method he used, still known as the Socratic method, was to engage interlocutors in rational dialogue and let them discover for themselves how questions and answers inescapably lead to the correct conclusions. One result of this didactic method was that pupils learned a great deal in a very efficient manner. Another was that Socrates never wrote anything down. Socrates and his students hesitated to commit their thoughts to papyrus. Static words were useful for the communication of information, they believed, but unsuitable when expressing one's deepest thoughts. Furthermore, written documents would simply expose the author to the envy and criticism of others. Had not Plato put his master's words into written form after his death, posterity would maybe never have known about this great man. In order to remain as true to Socrates' spoken words as possible, Plato presented his teachings in the form of dialogues. A wise man, Socrates himself would usually lead interlocutors through dialectic inquiry toward the inescapable truth.

In 399 BC, when Plato was twenty-eight years old, his revered teacher Socrates was put on trial. Charges brought against him were the spread of atheism and the corruption of youth. The authorities did not look kindly on Socrates' activities because letting young men think for themselves could become dangerous to the powers that be. Socrates mounted a spirited defense, putting his accusers to shame with his sarcasm and subtle irony. But his fate had already been sealed. When the moment arrived to pronounce the verdict, 280 of the 501 members of the jury voted that he should die for his transgressions. Plato attended the proceedings in court and later wrote an account of Socrates' defense. But when the time came for the condemned philosopher to die by drinking a beaker of hemlock, Plato was not present, claiming an indisposition. (This did not prevent him from describing in minute detail how the poison took over Socrates' body.)

Plato despised democracy. However, he was thinking of a different kind of democracy than the one to which we are accustomed. After all, the Athenian form of government was a democracy, and it was in this political environment that Socrates was condemned to death—in a proper court of law by a solid majority of jurors in a valid vote. How could so obvious a travesty of justice have come about? Something must have been wrong with the system. Obviously, to Plato at least, the regular folks were not fit to rule and to dispense justice. Hence, *democracy*, the power of the people (*demos*), was an inferior form of government. Thoroughly disgusted with the prevailing regime, a disillusioned Plato went to work, seeking a better form of judicial courts and government. The result of his inquiry was his seminal work *Politeia*, which, in Latin, became *De Re Publica*, and has been translated into English as the *Republic*. It was the world's first treatise on political philosophy and would inspire students of government throughout the next two and a half millennia. However, his ideas on proper administration were not quite thought through yet. For example, voting and elections were nowhere mentioned in the *Republic* (see additional reading).

Several attempts to put theory into practice by acting as advisor to dictators and tyrants were unsuccessful (see the biographical appendix). Disappointed and thoroughly dispirited, Plato set about revising his theory. Apparently the theoretical state as envisaged in *The Republic* was less feasible than he had hoped. If Plato wanted his teachings to have an impact, a major overhaul was required. His last manuscript, unfinished by the time of his death at age eighty, in 347 BC, was titled *Laws*. Containing twelve books, it was to be his longest and most practical piece of work. It was here that Plato gave a more realistic, if still utopian, theory of government. This time he realized that selection processes could not be avoided, and he discussed voting and elections at length.

The *Laws* recounts a conversation among three men walking on the island of Crete. They are on a pilgrimage to the temple of Zeus, and the conversation takes place over a time span of a full day. It is a long walk and they stop from time to time at shady places under lofty trees, all the time chatting away. One of the men, Megillus, hails from Sparta. Another, Cleinias had been sent from the town of Cnossos to found a new colony, Magnesia, on a remote part of the island and would like some advice on how to organize this city. The third man, the wise "Athenian stranger"— Socrates or Plato himself—is only too happy to oblige. He expounds on the social structure, urban design, and laws that should be introduced in the new colony. The description of the exchange as a conversation or dialogue—trialogue would be more apt—is somewhat of an overstatement. Plato reduces Cleinias and Megillus to uttering "of course," "that is very true," "by Zeus," and "OK" from time to time. (Well maybe not "OK," but something like that.)

The Athenian stranger's first advice is that the city should be composed of exactly 5,040 households. An average household—husband and wife with two or three children, an elderly relative or two to take care of, and a few slaves—would comprise about ten people. Hence the ideal city-state would count about 50,000 inhabitants. Why 5,040 households? The Athenian stranger asserts that this is a "convenient number." And so it is. It can be divided by all natural numbers up to ten and also by twelve, fourteen, fifteen, sixteen, and by a host of other numbers. Altogether it has fifty-nine divisors. This comes in very handy, the Athenian stranger asserts, when the need arises to partition the population and to allocate wealth or chores—for example when divvying up spoils or imposing taxes. Of course, immigration and emigration would have to be strictly controlled to keep the number of households unchanged. (Plato says nothing about what happens when the children of one household grow up and want to found their own household.)

In the center of the city, the citizens would erect the Acropolis with temples dedicated to Zeus, Hestia, and Athene. A wall would be built around the Acropolis, and outside this perimeter twelve neighborhoods would radiate outward, like pizza slices without the tip. Twelve tribes, with 420 households each, would settle in the slices. (Here, Plato may have taken a page out of the Israelites' narrative.) Each of the 420 households would be allocated two plots of land within its slice. One, near the Acropolis, would serve as a dwelling, the other, near the periphery, would be used for agriculture. If a household's place of residency were close to the center of the city, and therefore desirable, its outer plot would be farther toward the periphery and vice versa. If the agricultural parcel of land yielded little, it would be large, if it were fertile, it would be small. Everything would be allocated in a precise and mathematically fair way. However, the plots would only be leased to the households, remaining the property of the city for eternity. The "owners" would not be allowed to aggregate, subdivide, or sell them. Does not Plato's virtual state remind us of SimCity©, the enormously popular computer game that allows players to design a city according to their every whim?

Wealth, though permissible, would be strictly controlled. A household's minimum property would be the two plots of land initially allocated to it. They were meant to sustain the household. Anything below that would be considered insufficient and no ruler would allow a household's wealth to sink below the poverty level. Yet, by judicious trade, superior abilities, or sheer luck, some households would gain additional wealth. They would be permitted to hold up to four times the amount of their poorest compatriots. Every citizen's possessions would be meticulously recorded by the authorities, and taxes would be paid accordingly. The households would be divided into four wealth classes. Anybody found to be in possession of more than the maximum permitted amount, or of property that was not properly declared, would be obliged to surrender the surplus to the state. In addition he would have to pay a fine that would be used partly as a re-

4

ward to the good citizen who informed on the cheater. Plato envisioned both property taxes and income taxes, with the state deciding in each case which of the two would be more advantageous. The collected taxes would be used to pay for administrative expenses, military campaigns, the building of temples, and common meals.

Family life would have to be strictly regulated because a ruler "who imagines that he can . . . leave the private life of citizens wholly to take care of itself . . . is making a great mistake." After all, the Athenian stranger remarks, "among men all things depend upon three wants and desires . . . first, eating, secondly, drinking, thirdly, the excitement of love." If the third desire were not reined in, mayhem would most certainly result. Men and women were to marry and produce "the best and fairest specimen of children which they can" not because they want to, but as a duty to the state. And whereas the law should let alone matters of marriage, "every man should seek not the marriage which is most pleasing to himself, but the one which is most beneficial to the state." The best age for girls to get married is between sixteen and twenty, for men it is between thirty and thirty-five. Any man unmarried by the age of thirty-five would have to pay a yearly bachelor's fine "in order that he may not imagine his celibacy to bring ease and profit to him." (This is less ridiculous than it sounds. Nowadays, married couples and families with many children receive tax breaks, which, after all, are tantamount to taxing bachelors.) Couples that remain childless after ten years of marriage must divorce. If they do not, gentle, or less gentle, persuasion should be used to convince them to follow the laws of the state in this regard.

Once the norms and forms of behavior are decided, the questions of who would manage the city and how the administrators are to be chosen become relevant. In contrast to the *Republic*, where these matters were largely ignored, the questions received detailed treatment in the *Laws*. It is in Book VI that Plato lets the Athenian stranger make his first reference to the choice of civil servants: "And now, having made an end of the preliminaries, we will proceed to the appointment of magistrates."

The legislative, executive, and judicial powers that existed in Athens at the time consisted of three institutions: the Assembly, the Council, and the Courts of Justice. Important business matters like the issuance of decrees, the election of important officials, and the adoption of laws were made in the Assembly. It met ten times a year; in later times the frequency was increased to forty times a year. Every citizen in good standing—male, over twenty years of age, without debts to the state—was entitled to participate, and the number of those present often went into the thousands. When decisions had to be made, like going to war or granting citizenship to an out-of-towner, the attendees were polled. Votes were effected by a show of hands, and a simple majority sufficed to decide any issue. Since so many citizens were often present, the number of raised hands was just estimated.

Since decisions made by the Assembly were assumed to express the will of the people they were not subject to review by any higher authority. By definition, the Assembly was infallible. If erroneous decisions occurred nevertheless, it could only be because the citizens had been misled. Clearly Athens' confidence in the citizens' omniscience and infallibility was a far cry from Plato's *Republic* with its total denial of the simple people's ability to think and decide.

Less important and with less power than the Assembly, but nonetheless indispensable, was the Council of 500. Its task was to prepare legislation. No proposal could be put before the Assembly if it had not first undergone a preliminary screening by the Council. Thus, this institution had an important role in setting the agenda for the Assembly. The Council's 500 members were chosen yearly by lottery. Thus they were selected for the service not by their compatriots but by the gods. Members served for a single year but could be chosen for one more term during their lifetime.

The Courts of Law were the main instrument that guaranteed the orderly social functioning of the city. Juries, composed of at least 201 men for private lawsuits and at least 501 for public lawsuits, were selected by lot from a pool of 6,000 jurors who, themselves, had been selected by lot. The cases that came before the Court, like the one that condemned Socrates to death, were considered weightier than the day-to-day agenda items in the Assembly, and the jurors had to be more serious than assemblymen. They were required to be at least thirty years old. In addition, before they heard a case, they had to take an oath to adjudicate honestly. To allow poor citizens to take part in the administration of justice, jurors were paid for judicial duty. There was no presiding judge at the court sessions; in fact, there was no presiding anybody. The proceedings were predictably chaotic. But however boisterous the sessions were, being the voice of the people, Courts could not err. Just as with silly decisions by the Assembly, miscarriages of justice occurred only if the jurors were misled.

Such were the major institutions of the city-state of Athens. As for micromanagement, about a thousand civil servants were appointed every year. Since the danger existed that officeholders would abuse their position of authority in order to amass wealth or power, the prime aim in the choice of officials was to avoid corruption. Competence of any sort was not a prerequisite for the job; hence, there was no need to choose the best-qualified person. This is unfortunate because, the people being infallible, certainly the citizen most suited for the position would have been elected. As it was, city officials were chosen by lot.

To briefly summarize, it seems that everybody who had any sort of interest in running the city could either participate in the Assembly as he pleased, or was selected by lot, as in the Council, the Court, or the civil service. Votes were only taken in order to pass or reject laws, or to decide on the verdict in a criminal case.

But a select few officials did come to their positions by election. They were the ones whose jobs required special skills: warfare and money management. On the one hand, the ten generals who were elected and could be reelected every year needed experience and expert knowledge. On the other hand, the treasurers had to be wealthy in addition to being savvy, so that public money that they lost, due to mismanagement or through corruption, could be recovered from their personal property. These public officials were elected by a majority vote of the Assembly. We already know that the assembled citizens cannot err. So if an elected general lost a battle it must have been due to his having deceived the citizens. Upon his return he faced arrest, trial, and possible execution. Treasurers whose accounts did not add up must also have led the Assembly astray. They had to pay the missing amounts out of their pockets. After that they were sometimes executed nevertheless. This happened at least once. On that occasion, nine of ten treasurers were executed, one by one, until an accounting error was found just before the last one was to meet his fate.

Plato was not happy with this state of affairs. It was not so much the executions that bothered him. Rather, he objected to the fact that poor, uneducated masses could end up terrorizing the rich. Any dimwit was allowed to participate in the Assembly, and even though the members of the Council and of the Courts had to be older and presumably wiser, semiliterate bozos could be chosen by the lottery. How could a collection of such people make intelligent, informed decisions? Plato's *Laws* were meant to prevent the alleged mistakes that had been made in Athens from being made all over again in Magnesia, the colony to be founded on the island of Crete. As we shall see below, Plato tended to equate "rich" with "better educated."

In the guise of the Athenian stranger, Plato lays down his version of the best way to assign qualified individuals to jobs. After all, if unsuitable people would be appointed to public office, even the best laws would become useless. So, first of all, those who are to elect the magistrates, judges, and administrators would have to be well educated and properly trained in law. Only such accomplished electors would be able to make correct judgments, Plato states. The exclusion of the uneducated would prevent them from making inevitable errors. Second, candidates who wanted to run for office would have to give "satisfactory proof of what they are, from youth upward until the time of election." But not only the candidates' past would undergo scrutiny, their family history would also be subject to inquiry. Misconduct on the part of any family member, living or deceased, could be grounds for disqualification.

The election procedures that Plato proposes—there are many different variants, as we shall see immediately—are usually multistage processes. The early stage is designed to ferret out the obviously unsuitable contenders, in the subsequent stages the electors gradually advance to the most suitable candidates. Thus, blunders that were made at the start of the process could be corrected later on.

Most important for the survival and orderly functioning of the city were the guardians of the law, who would have to be chosen first of all and with the greatest of care. These esteemed personalities would be at least fifty years old and could serve for at most twenty years. At age seventy, "if they live that long," they would have to step down. (The Athenian stranger adds the helpful, if superfluous remark that a guardian of the law who is elected at age sixty would be able to serve for at most ten years.) Plato's scheme did not stipulate the separation of powers. Responsible for law and order in the city, the guardians' duties would include the enactment of laws, the administration of justice, and the registration of citizens and their wealth. As the work of legislation in the new city progresses and new laws are enacted, the guardians would be assigned further tasks.

Thus the guardians of the law would be the guarantors of justice and stability. Indispensable for the new colony's survival, their choice would have to be undertaken with particular care. As creators and founders of the new city, the burghers of Cnossos had a moral duty to see the fledgling colony through its first, still shaky period. Thus, the body of guardians should, by Plato's design, be made up of representatives both from Cnossos and from the new city.

The number of guardians would have to be odd so that close decisions would not end in a draw. And the settlers, who have a greater stake in the future of the city, should represent a majority in the legislative/judicial body. At this point, without further ado, the Athenian stranger declares at this point that the guardians of the law would number nineteen settlers and eighteen Cnossians for a total of thirty-seven. Why thirty-seven? No justification is given by the philosopher for this particular number except that it is odd. What if the eighteen chosen Cnossians were unwilling to leave their comfortable homes in order to move to an inhospitable colony? In this case, Plato asserts, it would be permissible to use "a bit of physical pressure" to persuade them to go.

While vague about the reason for choosing exactly this number of guardians, Plato was much more specific about the manner in which they should be chosen. He proposed a three-stage procedure during which the number of candidates would be whittled down successively to three hundred, then to one hundred, and finally to thirty-seven.

Since every soldier receives an education, all citizens who are or have been in the military would be qualified to take part in the election. By the way, women were considered suitable for army service in Plato's system and would therefore not be excluded from participation in the elections. Only dimwits, who had not been fit for the military, would be excluded. The election was to take place in a temple with the ballots deposited on the altar of the god.

Anybody could indicate a choice and vote for a candidate by writing the name of the preferred candidate, his father's name, the tribe to which he belongs, and the borough in which he lives onto a tablet and depositing it on the altar. Voting was by no means secret; the elector had to include his own particulars on the same tablet. Anybody who took exception with a particular tablet—because he objected either to the candidate or to the sponsor—was entitled to remove it from the altar. It would be exhibited for at least thirty days in the Agora, the marketplace, for everyone to see. If there were no objections to the objection, the rejected candidate's name would be permanently removed. Thus, any citizen could exercise veto power over any candidate he deemed unfit for this high office. Once all votes were cast, magistrates would count the tablets and announce the names of the three hundred top-ranked candidates.

In a second round of voting, the citizens would choose one hundred candidates from the reduced pool in the same manner. Finally, in the third round, the thirty-seven guardians of the law would be chosen from this shortlist. But at this point the Athenian stranger imposes a significant obligation on the voters. Before casting the third and deciding ballot, electors are required to solemnly "walk among the sacrificial animals." This innocuous requirement, seemingly meant to make the voters aware of the gravity of their choice and to invoke the gods' help in deciding correctly, actually limits the electorate. After all, who are the people with sufficient money to pay for sacrificial animals and with enough leisure to spend another day on an election? They are the wealthy citizens. Plato biases the election ever so subtly toward the rich who—he assumes not entirely without reason—are the better educated.

Having got this far, the Athenian stranger and his two interlocutors suddenly become aware of a problem. Nearly as an afterthought it occurs to them that elections require supervision. Even for the very first election, it takes magistrates to elect magistrates. Like the vexing question of the chicken and the egg, it was not clear how the process could start in a brand new colony. The problem was even more pressing, Plato asserts, because, as the proverb went, "a good beginning is half the business." Adding that, in his opinion, a good beginning is a great deal more than half the business, the Athenian stranger suggests a rather uninspired solution to jump-start the process. Upon their arrival at the proposed colony, a hundred Cnossians and a hundred settlers, the eldest and the best from among the two groups, would simply appoint the thirty-seven initial guardians of the law. After verification that the chosen magistrates were indeed qualified, the eighty-two Cnossians who were not appointed would be free to return home, and the settlers, together with the eighteen chosen Cnossians, would be left to fend for themselves. The suggestion raises more questions than it answers. While the determination of the oldest representatives would not be difficult, how does one determine the best? And once they have been determined, how do they appoint the thirty-seven? Leaving this question unanswered, Plato simply suggests that the Cnossians and the settlers choose the two hundred electors as best they can who will then do the appointment.

With the guardians of the law duly appointed, the city proceeds to the election of less crucial, but still essential, officeholders. First there are the higher charges of the military—the generals, brigadiers, and colonels. Candidates for generals, natives of the city, whose backgrounds have been examined and who have been found to be suitable for the post, are proposed by the guardians. Anyone who disagrees with their choices and believes that a certain candidate is unqualified may propose an alternative in his place. A primary election will then be held between the two, with the winner being admitted to the next round. In the final, deciding round, the three candidates with the highest number of votes are appointed. The generals themselves will then propose twelve brigadiers, one from each of the twelve tribes. Counterproposals may be made, followed by primaries, vote taking, and decision. But while every present or former soldier who so wished is allowed to take part in the election of generals, participation in the appointment of brigadiers and other high-ranking officers would be limited to the members of the different branches of the army—light infantrymen, heavy infantrymen, archers, cavalry—that they were to command. Finally, the lower charges would simply be appointed by the generals.

Next, the Athenian Stranger discusses the appointment of magistrates to the Council. This institution, which would be in charge of the city's administrative affairs, was to number 360 members. The number is convenient, the Athenian stranger claims, because it is thirty times the number of tribes (twelve), and it is ninety times the number of property classes (four). Men from age thirty and women from age forty would be eligible for election. For the election of the magistrates, which would take place every year over a period of five days, Plato prescribes a two-phase procedure to which he adds an interesting twist. The proposal was a hybrid between a regular, two-stage election and a lottery.

The first stage of the election, which takes place during the first four days, identifies a pool of candidates from among which the successful magistrates will be chosen on day five. While the election of about thirty magistrates from each tribe was desirable but not mandatory, choosing ninety representatives from each property class was obligatory. So on day one, the candidates from the wealthiest class would be selected. Every citizen would be obliged, under pain of penalty, to participate in the election. The following day, candidates of the second wealth group would be selected in the same manner. On the third day, when the candidates from the third class would be selected, only the members of the top three groups would be compelled to vote. The poor could cast their ballots if they chose, but were not obliged to do so. On the fourth day, when the candidates from the poorest class would be chosen, only the members of the top two wealth classes were obliged to participate.

Why does Plato propose such a convoluted procedure? Again, the aim is to give the better-educated citizens, that is, the wealthy property owners, a greater say in the composition of the Council. On the one hand, the rich will participate in all four rounds in order to avoid the fines. On the other hand, the poor, having already spent two days fulfilling their civic duties, cannot afford additional time away from their fields and cattle if they can help it. The result is that members of the two richer classes cast four ballots, members of the third class cast three, while the poorest only cast two. Note how the sly Athenian achieves his aim without making the poor feel cheated. He does not limit their right to vote but discourages their full participation in the election, all the while making them believe they are actually getting a break. Moreover, they are excused from voting precisely when it comes to electing the representatives from their own class. The rich decide who will be the candidates for the poor. Sycophants and yes-men stand a very good chance.

With the pool of candidates identified, it is now time to actually choose the magistrates. So on day five, it was everybody's turn again. From among the candidates of each wealth class, the citizens elect 180 men and women by majority vote. However, the procedure is not over yet because finally—this is the novelty of the procedure—half of the candidates, ninety from each wealth class, would be chosen for service on the Council by a lottery. Introducing an element of chance into the appointment, thus having God or Fortune make the final decision, permits more people to have a shot at governing and avoids discontent among the competitors. ("No hard feelings, it was God's choice.")

Plato not only limited the voice of the poor in the election; with the segmentation of the citizens into wealth classes he also curtailed their representation on the Council, albeit without their noticing it. By reserving for them the same number of slots as for the wealthy, the rich and the filthy rich, the poor were led to believe that they were equally represented on the Council. It is an economic fact of life, however, that every population contains many more poor than rich. Hence, the former would be quite underrepresented when compared to their numerical strength. The brilliance of Plato's scheme lies in the fact that the poor remained convinced of the exact opposite.

Next, the Athenian stranger discusses policing the city and the state. Like a boat that cannot be left without a captain, even for a short while, the city must at all times be controlled, Plato, a.k.a. the Athenian stranger, asserts. Streets, buildings, ports, fountains, temples, water supplies, and markets must constantly be inspected and regulated by the relevant officials. Some of them should be elected, others chosen by lot, and sometimes a mixture of both should be used to appoint them.

The first security service, akin to a police or sheriff department, would be made up of sixty men, five from each tribe, and of 144 deputies, twelve from each tribe. Plato does not specify whether the sheriffs and their deputies would be selected from each tribe by vote or by lot. Maybe not too many candidates registered for these positions anyway since life for two years in the open wilderness required an adventurous spirit on the part of the aspiring policeman, and the necessary equipment entailed considerable expense for the family. The groups would spend two months in each of the twelve different parts of the country. Their main task would be to make the citizens feel safe but they would also be responsible for keeping buildings intact, the irrigation working, the roads in repair, and the gymnasia in operation.

The town inspectors would see to it that building regulations are respected, structures are maintained, and water is of suitable quality. Six nominees from the two highest property classes would be elected, from which three are chosen by lot. They divide the twelve neighborhoods of the city into three boroughs and each takes one of them under his wings. The market inspectors maintain order in commerce and trading. They see to it that no injustices are being committed and, if commercial crimes or frauds do occur, that the perpetrators are suitably penalized. Ten candidates are elected by a show of hands from the top two wealth classes, from among which five are chosen by the lot.

As befits holy men and women, priests would not be elected by other mortals. Either their position is hereditary, in which case the question of election is moot, or they are to be chosen by divine chance, by lottery. After the priests, come the so-called interpreters. Their task would be to decipher the enigmatic messages of the Oracle of Delphi. As is appropriate for the inscrutable musings of an oracle, the passage describing the choice of the interpreters is one of the most enigmatic in Plato's *Laws*. "Thrice the four tribes are to elect four, each of whom is to come from among themselves; the three who receive the most votes are to be scrutinized; then the nine are to be sent to Delphi, where one is to be chosen from each triad," the Athenian stranger tells his listeners. Scholars have been puzzling over these words for centuries. Do the four tribes, voting as a body, select four persons, one from each constituent tribe, and then choose three of the four nominees? Or do the three groups of four tribes elect four nominees from any tribe in the group and then have three subsequent elections, in which all tribes participate, and in which each time three out of the four nominees are elected? Or does each elector cast four ballots for four nominees, with the top-ranked three going to Delphi? Or does each tribe separately elect four members from its own tribe, and then choose three of the sixteen? Exasperated, one scholar concludes his musings on the subject with the words "if I and others have misunderstood Plato, he has in this instance only himself to blame" (Saunders 1972). It is amazing how much ink has been spilled in attempts to guess what Plato really meant.

Judges of music and dance, conductors of the chorus, managers of schools and gymnasia are the only magistrates, with the possible exception of the generals, who are specifically required to have a certain expertise in the office they hold. For their appointment, only experience is to count; family background and probity of character are unimportant. Once a year, the citizens who are committed to such pursuits are obliged to participate in the election of these magistrates. In elections concerning music, ten candidates are nominated by a show of hands, one of whom is then selected by lot. For positions in gymnastics, twenty nominees are chosen from the second and third wealth classes; the poor and the filthy rich are excluded and the poor are even excused from participating in the election. Three are chosen by lot.

Lastly, a supervisor of education must be chosen. This magistrate's work is of supreme importance for the raising of a good citizenry. Since education of the city's youth is a vital concern this official is by far the most significant magistrate in the city. Great care must therefore be exercised in his appointment. He should be the city's best man (yes, this was one post for which Plato deemed women ineligible), at least fifty years of age, father of both sons and daughters, and of unblemished record. Since all the best citizens have already been appointed guardians of the law, there was no other way than to choose the supervisor of education from among them. He was to be elected for five years in a secret vote in which all magistrates participate, except the Council members.

Two details are noteworthy about the election of this most important of all officials. Firstly, only magistrates who have been vetted as competent individuals in a previous election are deemed capable of making such an important decision; hence, only this restricted circle of people participates in the appointment of the supervisor of education. This raises a problem. Since it is relatively easy to bribe a few individuals, the doors are opened to outside influence and corruption. Plato does not disregard the possibility of corruption even among magistrates, and in an attempt to avoid such aberrations he suggests, secondly, that the election of the supervisor of education be effected by secret ballot. In fact, it is the only one among the numerous elections in Plato's *Laws* that is secret. By the way, the desire to avoid situations where only a small number of corruptible electors take part in the appointment of a magistrate may be the reason why participation in most of the elections is mandatory.

Candidates for all positions, after selection by the lot or a show of hands but before appointment to the position, undergo rigorous scrutiny. During this assessment, his or her legitimate birth, flawless pedigree, impeccable reputation, lack of debts, and faultless character are publicly examined. If a candidate has not lived up to expectations, his or her selection is invalid and the appointment procedure must be repeated. The plight of at least one candidate is known who was nixed for a magistrate's position because he had not been good to his widowed mother.

At the end of the magistrate's term of office, an audit is held into the finances of the institution to which he belonged. The existence of a panel that would check all accounts, and the prospect of facing it, is meant to ensure that officials do not even think about enriching themselves at the city's expense. If temptation proves too great and the magistrate fails to keep his fingers out of the till, the panel would condemn him in a humiliating trial, leading to the payment of appropriate fines, in addition to the return of all he had taken.

The Athenian stranger then discusses the establishment of courts and the election of judges because "a city which has no regular courts of law would cease to exist." Whenever disputes arise among citizens, the court of first instance should constitute itself of friends and neighbors of the litigants who know what the dispute is about and can best adjudicate it. Occasionally the policemen, assisted in serious cases by their deputies, act as judges. In fact, every magistrate is in some way an adjudicator, the Athenian stranger asserts, since he must make decisions within the realm of his office and thus act as judge at such times.

If a plaintiff or defendant is unhappy with the judgment by the first court, he can move up the judicial hierarchy by appealing to a second instance, the tribal court, whose judges are selected by lot whenever the need arises. If one of the parties is still not satisfied with the verdict, an appeal can be lodged with the third and highest instance, the Supreme Court. The Chief Justices must be beyond reproach. In order to select them, the electors should also not be just any odd fellows but citizens of impeccable character, well versed in the laws. What better way to fulfill both requirements than by having magistrates choose the judges from among their own ranks? Thus, the Athenian stranger suggests, the members of each category of magistracy select one of their members—"the one who gives promise of rendering the best and most pious verdicts"—to serve on the Supreme Court. Consequently, this court will be composed of a policeman, a market regulator, a judge of music, a manager of gymnasia, and so on. The Supreme Court will render justice by a majority vote among the Chief Justices.

For the really serious cases, when a citizen is accused of having committed an injustice against the city, special tribunals must be constituted. Three high-ranking magistrates, chosen with the consent of the accused and the prosecutor, preside over the proceedings. (If the prosecutor and the accused cannot agree on the presiding magistrates, the Council decides.) They do not make the rulings, however; decisions are rendered by the assembled citizens who publicly vote for or against a guilty verdict. It is strange that after all the misgivings, Plato returns to the same institution that condemned his revered teacher Socrates to death.

The three men continue their conversation for hours, with the Athenian stranger expounding on nearly everything under the sun; family affairs, property laws, education, religion, food, sex, and many other subjects that make the social fabric of a city. Often the suggestions on how to run the city seem pulled out of a hat like a magician's rabbit, made up on the spot while enjoying the hike. Why ten judges of music and twenty of gymnastics, why use the lottery in this instance and not in another, why limit wealth to four times the poverty level and not to five? The stranger's proposals are good ideas but not necessarily the best. Nevertheless, the two listeners are in awe. Many suggestions sound brilliant to them on the basis of the Athenian stranger's authority alone.

Finally, they arrive at their destination and it is time to part company. Giving some last-minute pieces of advice, the Athenian stranger prepares to leave. But Plato cannot end the treatise without attributing a little praise to himself. Cleinias and Megillus are dispirited. They realize that without the stranger's help, they will never be able to make the new city flourish. Megillus has an idea: "Either we keep him with us and make him share in the founding of the city," he tells Cleinias, "or we give up the whole enterprise." "OK, so let's detain him," comes the answer.

And with this the dialogue ends.

ADDITIONAL READING

Republic

The first question Plato asked himself in the *Republic*, written about thirty years before *Laws*, was what justice is. He lets Socrates, the main character in the *Republic*, explore this question in the course of a long discussion with a circle of men. Cephalus ventures that justice consists simply in telling the truth and repaying one's debts. This is too simple-minded an answer and Socrates (that is, Plato) quickly counters with the example of returning borrowed weapons to a friend who has in the meantime gone mad. Surely it would be unjust to give him the means to kill himself? Polemarchus ventures that justice is doing good to friends and meting out punishment to one's enemies. But hurting enemies makes the punishers themselves unjust, Socrates points out, so that can't be the answer either. At that point Thrasymachus, a sophist who makes money dispensing philosophical advice, is unable to contain himself any longer. He blurts out a beguilingly simple answer: justice is what those in power decide it is. Now that really hit a bull's eye, and a full-blown argument erupts with thinly disguised, and sometimes undisguised, insults flying both ways. Finally, Socrates points out that a stupid ruler may enact legislation that is to his detriment. Would it then be just, by the ruler's own definition of justice, for citizens to follow these laws even though they result in the ruler's own removal? Hardly. Thrasymachus blushes and slinks away.

At one point, the dialogue veers off on a tangent when one interlocutor raises the question whether justice is a worthy aim at all, that should actually be strived for. If everybody else is just, maybe an unjust citizen could reap an advantage. Does injustice pay? (This argument anticipates twentieth-century game theory.) Socrates, never short of a counterexample, points out that even a gang of thieves, if they act unjustly toward each other, would not be a very successful gang. So justice, even among thieves, is somehow superior to total injustice.

Finally, the philosopher gives the answer his listeners had been waiting for. Justice means keeping a just order. Everybody should do what he does best and stay out of everybody else's business. If every citizen does what he is assigned to do, not because he is ordered to do it, but because he enjoys doing it, justice will reign. Citizens won't harm each other and the state will flourish because, on the one hand, justice leads to harmony and unity, while injustice, on the other hand, leads to sedition and revolution.

With this weighty question resolved, the next issue was how to organize the state in such a way that justice does, in fact, reign. As envisioned by Plato, the ideal republic would be sufficiently large to allow for an efficient division of labor, but small enough so that every citizen would have a personal stake in the state's affairs and take a vivid interest in running it. Everybody would have an assigned role and fulfill it to the best of his abilities. And what might that role be? Plato envisaged three kinds of citizens. (Slaves, even though they made up a sizeable part of every state's population, were excluded from consideration.)

First there would be the statesmen whom Plato calls the guardians of the state. They are the philosophers whose wisdom guarantees just and fair government. In order to prepare them for their task, they would undergo long and rigorous education, starting in childhood. Children and young people would not be allowed to read fiction because that would cloud their ability to think and argue rationally. After primary education and compulsory military service, ten years of instruction in mathematics would follow and another five years' training in dialectics. The by now thirty-five-year old aspiring guardians would then embark on fifteen-year apprenticeships in managing the affairs of the state. At age fifty, they would be ready to serve the state as philosopher-kings, making laws, adjudicating disputes, and dispensing justice. They would not own any personal wealth.

Then there would be the professional soldiers. The members of this class represent the police force and the army. Their task would be to preserve the existing order and to defend the state against foreign aggressors. Their defining attribute would be courage. These citizens would dedicate their lives to the community and, like the philosopher-kings, would not possess personal wealth. Housed, fed, and clothed by the state, they would not need to worry about material needs. Everything they require would be provided for by the state.

Wait a minute, does Plato advocate an early form of communism, more than two thousand years before Karl Marx wrote *Das Kapital*? Well nearly, but not quite. In contrast to Marx, Plato recognized that not everybody is ready to abandon the enjoyment of wealth, and he did not advise the general abolition of private property. Therefore he envisioned the third type of citizen.

This class, the largest of the three, would consist of everybody who is not part of the first two groups. With the administration and the defense of the state taken care of, these people would keep the economy going. They would produce, build, transport, and trade. Farmers and craftsmen would fall under this category, as would doctors, merchants, and sailors. These are the citizens who cannot do without private property. Plato allows them to own wealth, albeit in moderation. He determines the minimum amount of material goods needed to sustain a family. All wealth that surpasses four times that amount would be confiscated by the state.

Plato did not advocate a caste system. Assignment to any of the three classes would be by temperament, not by birth. Whichever of the three virtues—wisdom, courage, modera-

tion—was most developed in the child would determine his future path. The offspring of the third class could become guardians or soldiers while the progeny of the first two classes could become property owners. By the way, Plato made no distinction between men and women. Any citizen could attain any position in the state regardless of gender, and there could well be philosopher-queens.

Once society was suitably stratified, the question was what system of government would best suit it. Plato's preferred form was the aristocracy. Translated as "government by the best," it is not at all the feudal system of the European nobility of the Middle Ages. Peerages would not go from father to son regardless of whether the latter is an imbecile. Rather, according to Plato, aristocracy indicates a government by selfless philosopher-kings that would reconstitute itself afresh every generation. It was the best form of government that could be envisaged.

But even under aristocracy, danger lurked. Plato was well aware of the possibilities of corruption. He knew that not all soldiers would remain steadfast in the face of temptation. Especially war heroes who found honor (*timé* in Greek) in battle could be catapulted to the forefront, whence they would invariably turn on the philosopher-king in a military coup. The ensuing timocracy would be characterized by an overall aggressiveness toward the outside, and injustice toward the inside. Once in power, the former war heroes would certainly use their new position to amass fortunes, the result of which would be a plutocracy (*plutos*: wealth) in which the rich play top dogs. Now, wealth also produces poverty and by the nature of things there will be more poor people than rich people. One day, the latter will realize that they are more numerous and therefore mightier. The masses of simple people will overthrow the plutocrats and . . . institute a democracy.

Now don't you believe that this would be anything to look forward to. The people, unschooled und unsuited for administrative tasks, would make a horrible mess of things. Everybody would want to vote on matters in which they totally lacked experience and about which they had no knowledge whatsoever. Chaos would most certainly ensue. Surely democracy was not a viable form of government. Hence worse was to come. After a while, the baddest and boldest would take over and democracy, bad as it was, would metamorphose into something worse still: tyranny, the government of one.

Once the dead end of tyranny is reached, the only hope to escape it, according to Plato, is for the tyrant to take a philosopher to his side as advisor, or to become a philosopher-king himself, thus restarting the cycle. Very remote possibilities indeed. The self-assurance with which Plato predicts chains of events and their inevitable outcomes anticipates the certitude that characterized Karl Marx's description of social upheavals.

If aristocracy were the preferred form of government, how could the philosopher-king be chosen without going through the whole cycle of timocracy, plutocracy, democracy, tyranny to reach aristocracy? Plato abhorred involving the citizens in any decision process and, fortunately, in his system there was no need to do so. In the ideal state, governors would be selected according to their abilities and not because of popular preference for one person over another. As Socrates asserts, the qualities necessary to become philosopher-king—quick intelligence, memory, sagacity, ingenuity, fearlessness, and steadfastness—do not often grow together. Individuals who possess all these qualities are so rare that the state will hardly ever find more than one who fits the job description. Thus, elections and votes are superfluous.

BIOGRAPHICAL APPENDIX

Plato

When Plato was about forty years old, he traveled to Crete, Egypt, Cyrene, and to Syracuse. In the latter, on the island of Sicily, Dionysius the Elder ruled with an iron fist. The tyrant's brother-in-law, the philosopher Dion, enlisted Plato in an attempt to moderate the cruel regime. Together they tried to teach Dionysius the basics of a government based on philosophy, but to no avail. Worse, the angry tyrant reduced Plato to slavery.

He was barely rescued by one of his followers and only just made it back to Athens.

There he founded the Academy. It was the world's first university of sorts, where Plato taught his disciples about astronomy, biology, metaphysics, aesthetics, ethics, geometry, rhetoric, and politics. (One up-and-coming student and later teacher at the Academy was a young man by the name of Aristotle.) The Academy operated for a thousand years and was shut down only in AD 529 by the Roman emperor Justinian I, who claimed that it posed a threat to Christianity.

In 367 BC Dionysius died. Possibly his demise was helped along by doctors who poisoned him at the instigation of his son, Dionysius II, who could not wait to succeed him. Unfortunately, the elder Dionysius had been so busy ruling that he had not only failed to notice his son's ambitions, but, unfortunately, also neglected his education. The thirty-year-old prince, known more for his taste for debauchery than for his leadership qualities, was quite unprepared to take control. Again it was Dion who tried to remedy the situation. What the young man needed was a crash course in leadership and proper management techniques, and who would be better suited to instruct him than his old friend Plato? Recalling the failed experience with Dionysius's father, Plato balked at the suggestion at first and politely declined. Eventually he relented. It was,

after all, a good opportunity to put his teachings to the test.

However, it was not to be a successful experiment either. Dionysius II, jealous of his more capable uncle, sent Dion into exile. Plato himself was ill prepared for the intrigues at the court of Syracuse and with his friend gone, he remained without a protector. This was not an enviable situation for a sixty-year-old philosopher to be in. Plato took the wise course of action and left Syracuse. Back in Athens, he returned to the Academy that he had founded twenty years earlier.

Six years later, Plato was again invited to Syracuse. But the incompetent despot had learned nothing in the intervening years and was not prepared to change his ways. So Plato left again, once more with the job left undone. Dion in the meantime had reached the conclusion that philosophy did not work after all, and decided to set things right in the old-fashioned way. Arriving on the shores of Sicily with a military force, he quickly took over. Dionysius, in Italy at the time, hurried back to Syracuse, but was defeated. Now it was Dion who took a liking to power and became a tyrant. He was not to enjoy his new status for long, however. Three years later, he was killed by agents of the philosopher and mathematician Callippus, who was himself killed the following year. Obviously, philosophy was not the laid-back profession it is today.

CHAPTER TWO
THE LETTER WRITER

Like the administrators in the ancient Greek cities, the civil servants of the Roman Empire were concerned with governing well and dispensing justice. The magistrate and state official Gaius Plinius Caecilius Secundus, generally known as Pliny the Younger, raised a profound question about the proper way to vote on a particular issue. Born in AD 61 or 62 in what is now the Italian city of Como, Pliny had lost his father, a landowner, while still a child and was brought up by his mother. The main influence on his education was his mother's brother, Pliny the Elder, a Roman naval commander and avid natural philosopher.

In the year AD 79 a catastrophe of unimaginable magnitude befell Campania, a densely populated region in southern Italy. Shortly after noontime on August 24, Mount Vesuvius, a volcano that had been active for ten thousands of years, erupted. It had a long history of eruptions, the largest one having occurred around 1800 BC, during the Bronze Age. Nevertheless, people never stopped resettling in the fertile coastal area around what is today the Bay of Naples. The eruption in AD 79 had been preceded for days by tremors and earthquakes, but these were common occurrences in Campania and no cause for concern or panic. When the disaster eventually struck, it caught everyone by surprise. Sixteen years earlier an earthquake had hit the region and repairs had still been underway when the new and far larger calamity struck. The bustling towns of Pompeii with many, if not most, of its 20,000 citizens, and Herculaneum with a population of 5,000, were buried under three meters of lava and ashes. The exact number of those killed by falling debris, fire, and poisonous fumes is not known. But even the Romans, accustomed to losses amounting to thousands in battles and wars, regarded the immense death toll as exceptional.

Pompeii lay buried for many centuries. Only in the eighteenth century did archaeologists discover its remains. Buildings and temples, coins and artifacts and, above all, bodies of people, horses, and dogs were excavated. It was a macabre sight. Ashes and pumice had kept the remains of the victims in the exact position they occupied when they died. Their last moments' agony became apparent to those who dug out their remains 1,700 years later. Nowadays, the ruins of public buildings, villas, even brothels can be admired—except for some artifacts and erotic frescoes of the latter that are modestly kept hidden away.

Toward the end of the first century, the senator and historian Cornelius Tacitus compiled a history of the Roman Empire. Pliny the Younger had been an eyewitness, and Tacitus asked him to recount the tragic events. It was in honor of and in remembrance of his beloved uncle, Pliny the Elder, that he complied with Tacitus's request by writing two letters in which he described the occurrence. Eighteen years old at the time of the eruption, he remembered it vividly even in old age. His minute-by-minute account of the harrowing experience—as retold in the appendix to this chapter—is a chilling reminder of Nature's might and the human tragedy it may entail.

A year after the disaster at Mount Vesuvius, Pliny the Younger married for the first time—altogether he would tie the knot three times—and started appearing in court. He eventually became well known as a prosecutor as well as a defender of public officials. He began his career—typical for Roman noblemen—as a military officer in today's Syria, and then continued up the hierarchy of the civil service. Well versed in financial matters, he also served as the administrator of the state fund for veterans. His last position, in the year AD 111, was the posting as governor of the province Bithynia-Pontus, in today's Turkey. It is believed that he died a sudden death two years later, but nothing exact is known. The only indication concerning his demise is the sudden ending of the hitherto copious flow of letters. Throughout his career, Pliny was considered an honest and able functionary. The fact that he survived three emperors of disparate characters attests to his social and diplomatic skills.

Pliny the Younger was the first man known to have raised a question about the appropriate manner to count votes when making decisions. Nowadays, he is remembered mainly for his numerous letters. In fact, he developed letter writing into an art form. His style was such that it lent itself easily for publication. In fact, publishing his letters seems to have been the author's intent since he himself arranged most of them for publication. This makes for a certain lack of spontaneity, and one can feel that Pliny addressed not only the intended recipients of his epistles but also a broader audience. Through his abundant correspondence (altogether more than three hundred letters have survived) we learn about the day-to-day life in the Roman Empire during the first century, and of the concerns of a magistrate and member of the upper class. Usually each letter was devoted to a separate question. One letter, in particular, is the reason for our interest in Pliny, as we shall see below.

Some of the letters are surprisingly topical even today. Take the letter to his friend Julius Valerianus, in which Pliny tells about the problems simple people may have with their advocates. In Vicetia, today's Vicenza, sixty kilometers west of Venice, a former magistrate by the name of Sollers wanted to establish a market on his property. He asked the Senate for permission to rezone his real estate, but the townspeople objected. To represent them at the Senate, they hired the services of one Tuscilius Nominatus, paying him 6,000 sesterces and 1,000 dinars, which corresponded to what eight simple soldiers were paid for a year's service. Nominatus came to the Senate on the appropriate day, but for some reason the hearing was postponed. When the new date arrived, Nominatus was nowhere to be found. Not surprisingly, the people of Vicetia were angry. They had paid the agreed-upon fee but were now left with nobody to represent them. The original affair of the magistrate and his market had by then taken a back seat to the question of whether the townsmen had been tricked by their legal counsel. Had he pocketed their money and then left them high and dry? With this, Pliny's letter ends, leaving Valerianus in suspense. Pliny was not going to divulge the end of the story unless Valerianus either asked very nicely for the sequel or came to Rome in person to hear for himself the denouement.

Apparently Valerianus did ask nicely, because in the follow-up, Pliny recounts the remainder of the story. Nominatus had been summoned to a hearing to justify his behavior. His defense strategy was nothing if not ingenious. It was not disloyalty that kept him from arguing his clients' case, he claimed, but timidity. In fact, he had shown up for the hearing—there were witnesses to prove it—but then had been persuaded by friends not to oppose Sollers's plans. After all, Sollers was a member of the Senate, and Nominatus would be well advised not to oppose him, all the more so since the matter of rezoning had become a power play. Sollers was adamant about erecting a market on his real estate, if only to show that he wielded influence and occupied an important position. If Nominatus would persist in arguing the townspeople's case, his friends warned him, he would most surely come in for ill-will not only by Sollers but also by his fellow senators. A first indication of what would likely follow was that he had been hissed at by some senators at the first, aborted, hearing. So he had chosen to follow the safest course of action by leaving the chamber. Some argument!

Surprisingly, it worked. Nominatus, a skilled orator, spoke very convincingly. Shedding a tear or two at the appropriate moment, he asked for forgiveness without trying to defend his behavior. He found a sympathetic ear in a senator, and former consul, by the name of Afranius Dexter. Although Nominatus's actions had certainly been unbecoming a senator, Dexter said, they were not fraudulent. Why not simply order him to return the collected fees and acquit him? The suggestion was acceptable to most present. Everybody understood that a counsel could not be expected to make an enemy out of a powerful senator, so let the townsfolk of Vicetia fend for themselves. Only a certain Fabius Aper objected. Taking a surprisingly modern stand, he vehemently opposed the deal and insisted that Nominatus should be disbarred for a period of five years. He was immediately supported by a tribune of the plebs, a sort of public defender of the simple folks, who was only too happy to lash out at the legal establishment. Counsels were routinely bought and sold, he cried out, they bring litigation in order to create work for each other, and they conspire and sell their clients' cases. And instead of being satisfied with the honor that goes with the reputation of being successful, they collect large fees for their alleged services. The scathing attack was to no avail, however, and the honorable senators—how could it have been otherwise?—agreed with Dexter. Nominatus would simply return the fees and everyone would get on with life.

Thus ends Pliny's letter to Valerianus, except for a postscript. The tribune's public lament did have an unexpected effect. The emperor decreed that it would henceforth be forbidden for counsels to solicit fees. Now that's a legal system to our liking. It is not without satisfaction that Pliny remarks to his friend that he himself had always made a point of refusing gifts or even acknowledgment for his services as counsel. Caustically he also remarked that once his colleagues find out that they would have to behave in the future as he had always done, he would not be their dearest friend.

Now we come to Pliny's preoccupation with voting procedures. It was this same Afranius Dexter, or rather his mysterious death, that would trigger Pliny's interest in the subject. On June 24, AD 105, the senator's lifeless body was found in his home. The circumstances of his demise are unknown. All that is certain is that he met a violent death; otherwise, there would be nothing to tell. Did Dexter commit suicide? Or, not sufficiently strong-willed to kill himself, did he order someone to do the job? The authorities apparently carried out a crime scene investigation but we do not know the details. Pliny's letter does not reveal where Dexter's body was discovered and by whom, nor what device was used in the killing, nor if and where the instrument was found.

Suspicion immediately fell on Dexter's slaves. They were the only ones who could have carried out the murder if, indeed, it was one. The case went to the Senate. Cognizant of all known details, the senators had to decide on the verdict and the sentence. If guilty, the slaves would—at best—be banished to an island or—at worst—be executed; if innocent, they would be freed. The verdict and sentence were not separate. By pronouncing a sentence, the senators implied the verdict. Banishment being less of a punishment than execution, the latter would imply some partial guilt or maybe the existence of extenuating circumstances; for example, Dexter could have asked a slave to kill him. The verdict then could be guilty, guilty with extenuating circumstances, or innocent. The concomitant sentences would be execution, banishment, or, the slaves could be released.

The problem was that the decision at hand was not binary—guilt or innocence—but ternary—execution, banishment, or release. By allowing three options, the way was opened to various machinations and manipulations. And Pliny, advocating a somewhat dubious course of action, as we shall see presently, was foremost among the manipulators. However, after everything was over and done with, the matter did not let him rest and he was honest enough to admit that his actions may have been questionable. Troubled by pangs of conscience, he describes the course of events in a letter to Titus Aristo, a wise and learned friend to whom he often turned with legal questions. He wanted to hear Aristo's opinion about whether he had acted appropriately or whether he had made a mistake.

The problem Pliny raised was not one of private or public law, but rather one of procedure. At the start of the hearing, it became apparent that from among the three options, death, banishment and acquittal, a large majority, Pliny among it, favored acquittal. But it was no absolute majority; let us say that 40 percent of the senators favored acquittal. The proponents of the other two sentences, death penalty or banishment, were about evenly divided. Now, seeing that in a three-way vote the majority would opt for release of the slaves, the proponents of execution and of banishment decided to form a coalition. As was the custom, they got up from their seats, walked over to one side of the Senate, and sat down together as a group. Together the members of this large group—about 60 percent of those present—maintained that they favored banishment so that the slaves would not be acquitted.

But this was unfair, Pliny maintained. Even though the demarcation among the senators did follow the guilty versus innocent division, he claimed, the punishment of execution was about as far removed from banishment as banishment was from release. So it was entirely unreasonable for the proponents of banishment and of execution to form an alliance. If anything, it would have been more natural for the proponents of banishment to form a coalition with those who wanted to acquit, since both verdicts would let the slaves stay alive. But here they were, the proponents of execution and of banishment, united only in their common goal to thwart an acquittal. In order to achieve this goal, they were prepared to put off their disagreement with a temporary display of unity. Pliny, who presided over the assembly, was dismayed.

He let the senators know about his indignation. Even if they opposed acquittal, he cried out, it would be wrong to temporarily feign harmony when, in truth, there was no agreement about the verdict. He then decreed that each opinion would be counted separately and directed the two factions to break up their coalition. To justify his instruction, Pliny quotes the law that deals with voting procedures: "*Qui haec censetis, in hanc partem, qui alia omnia, in illam partem ite qua sentitis.*" Freely translated, this reads as "those who are of a certain opinion move to one location in the Senate, those of you who support all the other proposals move to the location which corresponds to how you feel about the issue." Desperately trying to substantiate his particular interpretation of the legal text, he subjects the sentence to a word-by-word analysis. Pliny stresses that the legislators wrote "those who support *all* the other proposals," which he interprets to mean that each of the other proposals must be supported on its own merit. He further supports his version by pointing to the phrase *qua sentitits*, which stands for "according to how you feel." Taking this to mean "*exactly* how you feel," this excludes your joining a group with a different opinion. Consequently, he orders the senators to split into three separate groups according to the options they truly favor, and to take their seats in three different areas of the Senate.

Still, Pliny's interpretation is somewhat tenuous, and the Latin text may allow another interpretation. *In illam partem ite* refers to the Senate area in the singular ("move to the location" not "move to the locations") and may therefore mean that *all* those with differing opinions are to move to *one* separate area in the Senate. Thus, the proponents of acquittal would sit and be counted as one group, and all others, that is, the proponents of execution as well as of banishment, would sit in another area and also be counted as a single group.

Since Pliny presided over the meeting he was the one to decide, and the senators shuffled dutifully toward the benches assigned to each of the three options. Of course, Pliny's instruction could not be enforced, because the senators' true beliefs could not be ascertained. It was precisely in order to prevent manipulations, like the building of coalitions, that voters were required in later times to swear that they would cast their ballots only according to their true intention. But no oaths were taken in the Dexter case and this, in the end, determined the outcome

Pliny was confident that in a three-way vote, a relative majority of 40 percent would favor acquittal, while only 30 percent each would opt either for capital punishment or for banishment. He had underestimated his opponent. The leader of the faction that favored execution realized that his proposal was headed toward defeat. In order to avert a debacle, he once again sided with those favoring banishment. When his friends saw him headed toward the side of the Senate where the proponents of banishment sat, they followed suit. As a result, in the revised procedure, about 60 percent voted for banishment, against 40 who voted for the slaves' release. With this the slaves' fate was sealed: they would not be executed but they would not be released either.

Ultimately, Pliny's attempts at manipulation had been to no avail. In the naive expectation that all senators would be sincere, he had manipulated the procedure so it would guarantee his preferred outcome. But he fell victim to a more sophisticated manipulator who had known from the out-set how to play the game. The proponent of capital punishment realized that siding with the banishers would at least guarantee him the slaves' relegation. Even though this was not his favored outcome, it was his sec-ond choice, and by hiding his true intention he managed to obtain it. Pliny and his faction also got their second choice: at least the slaves were not killed.

Pliny's opinion would have fared no better when translated into a mod-ern legal system. Today's courts first decide on the accused's guilt or in-nocence and later on the sentence. In the Afranius Dexter case, the fac-tion that believed the slaves guilty—those favoring execution as well as those favoring banishment, about 60 percent of those present—would have won in the first stage. (Pliny would have written a dissenting minor-ity opinion.) In the next stage, with extenuating circumstances and other facts and issues taken into account, about 70 percent of the judges, Pliny included, would have opted for banishment.

Was Pliny right in acting the way he did? Instead of two successive bi-nary decisions—decide on guilt or innocence and only then decide about the penalty, or decide on whether to let the slaves live or not and only then decide about freedom or exile—he had pushed through a ternary de-cision—acquittal, banishment, execution. In later times, choices involving more than two alternatives, or elections involving more than two candi-dates, often require an absolute majority, that is, a majority of at least 50 percent. Such a majority guarantees that the preferred option beats not only every alternative but also any combination of alternatives. In case no absolute majority is achieved, runoff elections are usually held between the two top-ranked alternatives or candidates.

In principle, Pliny's suggestion of a ternary choice, with only a relative majority required, is also acceptable, but only if it is agreed upon in ad-vance. It is certainly not acceptable to make up the procedures as one goes along. Insofar as Pliny attempted to model the voting procedure in such a way as to ascertain his preferred outcome, he was certainly in the wrong. The rules of the game must be specified and agreed upon at the outset, before any occasion to use, or abuse, them presents itself.

Unfortunately, we do not know how Titus Aristo felt about the matter. His answer has not been passed down to us.

ADDITIONAL READING

The Eruption of Mount Vesuvius

Pliny the Elder lived with his sister and her son in Misenum, about twenty kilometers from Mount Vesuvius. On the morning of this August day he sunbathed, then took a cold bath, then had lunch and was resting with his books when, at about two o'clock in the afternoon, his sister rushed in to tell him of a cloud of unusual size and appearance that rose above Vesuvius. In fact, what she saw was the eruption column, hot volcanic ash that, it is estimated today, reached forty kilometers up into the stratosphere. Pliny rose to get a better look. What he saw kindled his interest and he decided to take a boat to the other side of the bay in order to further investigate the matter. He invited his nephew to accompany him but the latter, to his good fortune, declined. Pliny the Younger preferred to stay behind and continue studying. At that moment a messenger arrived with an urgent letter from Rectina, the terrified wife of Tascius Pomponianus, whose villa lay at the foot of Vesuvius. She beseeched Pliny to save her, and he immediately ordered preparation of a larger vessel. He would rescue as many people as he could from what was clearly more than a curious cloud formation.

The boat commanded by Pliny held course directly toward Stabiae, on the other side of the bay. With the wind blowing from behind toward the southeast, the boat made good progress. All the while ash and lava particles rained onto the deck and into the sea, which had become shallow from the debris falling into the water. Pliny's helmsman urged him to return, but the courageous commander would have none of that. "Fortune helps the brave," he cried out and gave the fateful order "head for Pomponianus!" Rectina's unfortunate husband had already loaded his belongings onto a boat and was ready to flee. But the winds that helped Pliny's boat arrive, made it impossible to sail away from the coast and left him and his family stranded. Upon arrival, Pliny embraced the frightened man and comforted him, while flames loomed close by. The commander attempted to restore some calm in the terrified crowd. He explained that the houses of farmers who had left their homes in a panic without extinguishing their hearths must have caught fire. In order to lessen the others' fear and to show, or purport to show, his lack of concern, he took to the public baths where he went through his usual routine of bathing and dining and then going to sleep. He did not just pretend, because the snores of the corpulent man could be heard from outside the room. But during the night the tremors became so violent that buildings were in danger of coming loose from their foundations. Rocks continued to fall, and ashes and rubble rose so high that flight would soon become impossible. Pliny got up from his resting place and discussed with the others which course of action to take: stay inside and risk getting crushed by the collapsing building or go outside and risk getting caught in the rocks raining down.

They chose the latter course and with pillows tied to their heads for protection against the falling rocks started to make their way toward the beach. Ashes and fumes rendered the night pitch black and the group had to seek its way in total darkness. When they arrived at the shore, the winds were still against them and there could be no thought of departure. Plinius set down to rest, drank some water, and waited. Eventually, the flames and the smell of sulfur became so strong that everybody took flight. Supported by two slaves Pliny tried to rise, but—overcome by the fumes and the dust-laden air—immediately collapsed and died.

Back at Misenium, mother and son waited anxiously for news of Pliny. Taking a cue from his uncle, the younger Pliny sat on a terrace, pretending calmly to read a book, and even taking notes. Later he would admit that this display of nonchalance was not so much an expression of bravery but of the foolhardiness of youth. By the next morning the ground was shaking so much that carts rolled to and fro, even after stones had been wedged under the wheels. The water in the sea had receded and fish and other sea creatures were left high and dry on the sand. Frightening dark clouds massed together and fires broke out everywhere. A friend from Spain, who had been visiting, urged them to flee. "If your uncle is alive, he most certainly wants you to be safe; if he has perished he would want you to survive. So why are you waiting instead of fleeing?" And with this, not waiting for any response, he took off. Still, mother and son were reluctant to escape without news of their relative.

Shortly thereafter the cloud began descending all the way to the ground, and now Pliny's mother urged her son to flee. A man as young as he could make it to safety, while she, frail with age, would die happily, knowing that her son would survive. But Pliny would have none of that; he took his mother's hand and made her walk toward the outskirts of town. After a while, they turned off the street, so as not to be crushed by the fleeing crowd, and sat down. Soon total darkness descended. Parents, children, and spouses sought each other in total obscurity. Women screamed, children cried, men shouted. Eventually the darkness lifted somewhat but the light was no sign of relief. To the contrary, it was the fire that was drawing nearer that lit up the ghastly scene. Ashes and pumice continued raining down. Pliny was certain that he and his mother were going to perish and with them the world around them.

At last, after more than a day of dread and terror, the darkness started to dissolve. Soon it was no more than a cloud, then just a fog. Daylight appeared again, to reveal a changed world to the horrified survivors, "buried in ash like snow." Everything was destroyed; many survivors had gone mad during the ordeal. Nobody knew what to do and everybody feared what the future would bring. Eventually news came about the uncle's fate. The body of Pliny the Elder had been found by survivors, intact, as if he were asleep.

The death toll of the eruption of Mount Vesuvius is unknown but the devastation of the region was nearly complete. Only those who, like Pliny the Younger, took flight immediately, leaving all their possessions behind, had any chance of survival. The others were caught in the pyroclastic flow, a ground-hugging avalanche of blazing hot ash, pumice, rock fragments, and volcanic gas that had rushed down the side of Mount Vesuvius at a speed of hundreds of kilometers per hour. It is estimated that Vesuvius spewed out 4 billion cubic meters of rock.

The AD 79 eruption was no one-time disaster. Another major eruption occurred in 1631, leaving 4,000 people dead. And Mount Vesuvius is still active today. Nobody knows when the next calamity will take place and how devastating it will be. Today, about 3 million people live within the endangered radius.

THE MYSTIC

In the Athenian Assembly the choices to be made were usually of the form yes/no, for/against, guilty/innocent. The decision-making process usually worked all right, because deciding between two options presented no special problems; a simple majority vote does the trick. And whenever an official had to be chosen from among more than two candidates, the selection was usually deferred to chance, or god, by drawing lots. With time it became apparent, however, that choices between more than two alternatives were inevitable, and electors were often hard put to agree on a winner. Reluctant to let the lot—or fortune, or god—decide on important issues, many institutions drew up their own rules as they went along. By and by, special methods were instituted to elect emperors, popes, and the doges of Venice. But house rules did not always meet with everybody's approval. Pliny may have been the first to organize, and to manipulate, a three-way decision. It was only the earliest-known instance of problematic decision-making procedures, innumerably more followed. For example, thirteen centuries after Pliny, during the Papal Schism that lasted from 1378 until 1417, two, and at a certain time three, popes reigned over their flock. The need for more sophisticated electoral schemes became apparent during the Middle Ages.

Until quite recently most researchers believed that interest in the theory of voting and elections had started toward the end of the eighteenth century, at the time of the French Revolution. But toward the middle of the twentieth century, medievalists were surprised to discover manuscripts in the Vatican Library and elsewhere that showed that sophisticated ideas had already been around half a millennium earlier. As far as is now known, first mention of an electoral method other than the simple majority vote was made in the thirteenth century by the Spanish theologian and philosopher Ramon Llull.

Llull, also known as Raymond Llully or Raimundo Lulio, was born in Palma de Majorca, in about 1232. Today's scholars consider him one of the most influential intellectuals of the Middle Ages. He sprang from a well-to-do Catalan family. His father had aided King James I of Aragon in the conquest of the island of Majorca and received a valuable estate in return. It was only natural that the son of this loyal subject should enter the services of the king. He became seneschal, an administrative head of the royal household, at the Court of Prince James, who would later become James II, King of Majorca. But all of a sudden, at age thirty-three, Llull turned away from the worldly life at court to become a monk, missionary, and philosopher.

During his long life Llull wrote some 260 works on theology, philosophy, science, and mathematics. Actually, he did not take credit for everything he penned. On at least one occasion "the Lord suddenly illuminated his mind, giving him the form and method for writing the [best of all] books against the errors of the unbelievers," he wrote in his autobiography (referring to himself in the third person). This experience earned him the title of Doctor Illuminatus. His semi-mathematical proofs often relied on combinatorial techniques and on a logical system of his own invention, which he also applied to the search for moral and theological truths. By the way, this had a profound influence on the German thinker Gottfried Wilhelm von Leibniz. After studying Llull's writings, the great seventeenth-century philosopher was led to the conviction that sometime in the future, philosophers would settle their disputes like accountants do: with calculators. Leibniz's rival, Isaac Newton in England, also kept some of Llull's works on his bookshelves.

When faced with a choice among different alternatives, Llull liked to group the options into pairs and make two-way comparisons within each couple. (We will see presently that this technique of making decisions formed the basis for his proposals on the subject of elections.) In fact, the number two and all its powers were a source of constant fascination to Llull, centuries before the computer age in which the binary system became ubiquitous. In his attempts to binarize everything, he did not stop at the doors of the church and may have even sought to adapt the most basic tenet of the Christian faith, the Trinity, to the binary system. By adding Mary to the Father, the Son, and the Holy Spirit he is said to have tried to make the Holy Trinity into a Holy Quaternity or, in modern parlance, into a Holy 2^2. Such a revolutionary challenge to accepted principles did not sit well with the church administration, and he was soon whistled back into line. It is in part due to his fascination with the number two and its powers that Llull is considered one of the forefathers of computer science. The former, and failed, troubadour is also regarded as the father of Catalan literature. He wrote belles lettres, foremost among them the inspirational novel *Blanquerna*, about which I will have much more to say later.

In devising his contributions to the theory of voting, Llull was led by the conviction that divine truth—the one and only God-given answer to any question of choice—is always out there somewhere. All that electors have to do is grope their way toward this truth. If they are completely honest, they will spot the best candidate or the preferred alternative. But people are sinners, which prevents them from always recognizing the truth. Hence methods must be devised that allow the discovery of God's will in spite of the electors' deficiencies. It was Llull's objective to find such methods.

Choices between two alternatives and two-way comparisons between candidates lie at the heart of Llull's voting method. Among his numerous writings three texts include discussions about elections. One of them had been available throughout the centuries, but nobody had taken notice. The two others are fairly recent discoveries. The first source that I will discuss, the one that had always been known, is the devotional novel *Libre d'Evast e d'Aloma e de Blanquerna*, which Llull wrote in Montpellier in 1283. It contains something that is of particular interest to us, namely a chapter on elections.

The title refers to a young man by the name of Blanquerna, son of the wealthy Evast and the beautiful Aloma. The *Libre* is a thirteenth-century parenting guide, with Llull acting as a medieval Dr. Benjamin Spock. In the first chapters the author deals with the preferred feeding practices for a baby and with a child's upbringing. When Blanquerna comes of age, he decides that he wants to become a hermit. This does not please his mother who must have hoped for a flock of grandchildren. She conspires with a friend who has a similar problem with her daughter Natana, and the two young people are brought together. As planned, they fall in love and their families look forward to a wedding. No tragedy à la Romeo and Juliet looming here since everyone agrees that Blanquerna and Natana are a match made in heaven. So what is the problem? Well, heaven is all right with the two, but the match is not. Blanquerna and his girlfriend are as religious as they are in love. And as we would expect from such an inspirational novel, faith wins out over passion. The wedding is off and the young people devote the rest of their lives to the church.

They make their careers in the ecclesiastical hierarchy, and with time Natana becomes abbess of a monastery. Blanquerna is destined for even more greatness, but first his faith is severely tested. In one ordeal he happens upon a knight who is about to carry off a maiden against her will. The lady's cries for help are not lost on Blanquerna, and with only faith as his weapon the monk manages to wear the knight down with his rhetorical skills. Naturally the maiden is quite prepared to show her gratitude, but Blanquerna, firm in his vow of celibacy, delivers her home intact. The virtuous young man manages to withstand numerous other trials and tribulations and is eventually elected abbot of his monastery. Moving up the church ladder, Blanquerna rises to become bishop, then archbishop. The climb does not occur without a hitch, however. Just as he was about to be elected bishop, his enemy, the archdeacon, tells his supporters that Blanquerna will make them take vows of chastity, poverty, and obedience. An election takes place and since there are only two candidates, a simple majority suffices. Most electors vote for Blanquerna but a minority of the fine men, supporters of the archdeacon who do not cherish life with neither women nor money, refuse to accept the outcome. A dispute arises and both sides take their grievance to the pope, who decides—obviously—in Blanquerna's favor. Later on, the righteous Blanquerna is to become pope himself. But upon reaching the top Blanquerna has enough of it all and decides to fulfill a dream he had nurtured since he was a youth: he becomes a hermit.

It is chapter 24 of that novel that concerns us here. There Llull presents his electoral scheme. The chapter is titled *En qual manera Natana fu eleta a abadessa* (How Natana was elected abbess). The old mother superior of the monastery has died and a new leader has to be selected. Unhappy with the customary electoral process, Natana tells her fellow nuns about a new method that she had heard about. It was contained in the "Art of Finding the Truth," written by—who else—Ramon Llull. The method permits a search for "the sister who is most suitable and best to be our abbess" without allowing any possibility of error. Natana then describes the two-stage system to her sisters.

Before the procedure starts, each nun takes an oath to tell the truth. No swindling here, no vote buying and no collusion, everything is visible to the Almighty above and the nuns will vote exactly according to how they feel about their sisters. The first stage of the process consists of choosing an odd number of electors from among the convent's twenty nuns who will automatically be candidates for the top job. Natana suggests seven electors, giving no reason for this particular number except to say that it was better than five. She may have been influenced by mythology, culture, religion, or superstition, in which the number seven is deeply rooted: there are seven days of the week, seven wonders of the world, seven deadly sins, and seven virtues. Natana does not even say why the electors should be whittled down from the full house of twenty. Maybe she wants to reserve a more decisive say for better-qualified, smarter, or more deserving nuns. Maybe it is meant to make the second stage of the procedure, in which the rather tedious actual selection takes place, more manageable. Be that as it may, seven electors it was to be.

So the twenty sisters are asked which seven nuns are the most suitable to elect, and be elected to be, the superior. The seven nuns who get the most votes are the ones both to stand for election and to elect the abbess. But the seven sisters may feel that other nuns are equally deserving and should be added to the roster of candidates. They are therefore allowed to add additional sisters—Natana suggests two—to the list. There are now a total of nine candidates, seven of which also serve as electors. With these preliminaries out of the way, the process can take off. The eleven sisters who are neither candidates nor electors relax and watch the proceedings.

The distinctive feature of Llull's scheme manifests itself in the second stage. In contrast to traditional procedures the nine candidates do not stand for election together. Rather, each nun on the roster runs against each of the other nuns in a series of two-way comparisons. A tally sheet is prepared, in which the intermediate results are entered. Pairs of nuns present themselves to the electors who decide which of the two sisters is the more suitable. The winning nun in each showdown is awarded one point that is entered in the tally sheet into the cell bearing her name. Since there are nine nuns running for the job, each candidate runs eight times, once against each competitor. Altogether there are thirty-six pairings. (The first nun runs against eight competitors, the second against seven, and so forth. In total that gives $8 + 7 + 6 + 5 + 4 + 3 + 2 + 1 = 36$ pairings. For the more mathematically inclined: if the number of candidates is n, the number of pairs equals $n(n - 1)/2$.) After all pairings have been examined, the points are totaled and the sister "who has the most votes in the most cells is declared the winner." If a nun is so much superior to her eight competitors that she wins every showdown, she would receive a maximum of eight points. But she does not have to win all eight show-downs to become the new mother superior. It suffices that she simply gar-ners more points than any of her sisters.

There could be ties, of course. In fact there can be two kinds of ties, one of which Llull considers and another that he glosses over. The first one occurs if two or more sisters win the same number of two-way con-tests. In this case, says Llull, another election along the same lines is to be held among the tied nuns. What if some nuns are again tied? This possibil-ity, which could occur over and over again in a closely contested race, was not considered by Llull. Apparently he thought the likelihood of this happening would be too remote.

But another sort of tie can occur, namely in the two-way comparisons themselves. In the showdowns, when two candidates run against each other, both nuns could conceivably receive the same number of votes. In an attempt to avoid this possibility Llull had insisted on an odd number of electors, in the erroneous belief that this would guarantee a clear win-ner in each of the comparisons. But he was wrong. By allowing up to two additional nuns to run as candidates, he defeated the purpose of hav-ing an odd numbers of electors. When one of the seven electors is paired with one of the two additional candidates, the decision as to who is the winner is made by six electors. In this case the outcome could be 3:3 and the showdown would produce no winner. Llull does not discuss this possibility.

This annoying shortcoming did not seem to bother Llull and he de-clared—via Natana—that in the proposed electoral method no error could occur. The scheme was meant to reveal God's will. Since the winning can-didate had been compared to each of the other candidates and found to be superior more often than any of the other candidates, she is obviously the one God deemed most worthy of the new task. As pointed out above, the winner does not have to be superior in every single showdown. It suf-fices if she wins more duels than any of the other candidates. Thus the possibility exists that the nun who is eventually chosen to be abbess may have been considered less suited for the job in some of the two-way con-tests with other nuns.

Back to the novel. Of course, it came as it had to, and Natana was elected abbess. Actually, the title of the chapter already gives the outcome of the election away, but then again, Llull's purpose was not to recount who, but how, the new abbess was to be elected. The newly chosen mother superior felt a bit uncomfortable at having been elected by a procedure she herself had recommended. After all, the defeated candidates could harbor the suspicion that the procedure Natana herself had proposed had somehow favored her. In order to dispel all doubts, at least about the results if not about the method, Natana insisted on examining the outcomes of the thirty-six two-way comparisons and only after having found no error, was she persuaded to take on the job.

Before we move on, we have to deal with one more issue. From one somewhat unclear sentence in the novel the question arises whether Llull may have had a slightly different method in mind. When expounding on the election of the abbess Llull wrote in *Blanquerna* "*e sia elet aquel qui haurá mes veus en mes cambers*," which means, in English, "and let her be elected who has the most votes in most cells." Most votes in most cells? This is ambiguous. Should just the number of wins be registered, as we just described? Or did Llull mean that the votes each candidate receives from all electors in each of the showdowns should be counted?

To illustrate, let us say Sister Ana wins five contests with five votes to two, and loses three showdowns with three votes to four. Meanwhile, Sister Berta wins six of her showdowns with four votes to three, and loses two with one vote to six. If the number of wins is to be counted, Sister Berta becomes abbess with six wins against Ana's five. But if the votes for each candidate are the relevant criterion, Sister Ana is the winner with a total of thirty-four votes; Sister Berta loses, having garnered only twenty-six.

The confusion stems from the fact that *veus* can mean both "votes of the elector" and "winning points." The person who dealt with the issue first was a German philosopher and expert on Llull's manuscripts by the name of Martin Honecker. An avid medievalist and philologist, his enthusiasm and abilities unfortunately did not extend to quantitative matters and he got everything mixed up. In his explanation of Llull's text he wrote that the candidate "who amongst all pair-wise comparisons has totaled the most votes, that is, has won most of the individual comparisons, wins the election." Well, "totaling most votes" is quite different from "winning most comparisons." The first clause of Honecker's sentence does not imply the second. One would have expected a philosopher to be better versed in the interpretation of texts, even if they contain a little mathematics. But then again, there are many mathematicians with an insufficient grasp of philosophical matters.

Finally, Friedrich Pukelsheim, a professor in Augsburg, Germany, set things right. He maintained that it was not the votes that count but the wins. To make the case for "one win, one point" he called attention to the following sentence in *Blanquerna*: "Let it be written down [in each round] who has the most votes." Since only the *name* of the winner is written down in each showdown, he argued, and not the *number of votes* that the candidate received, it is the number of wins that decides who will become abbot, abbess, or bishop.

The second text by Llull, which presents the election method in a more detailed fashion, predates the novel *Blanquerna* by about eight to ten years. It is the tract *Artifitium Electionis Personarum* (Method for the election of persons). The sole known copy of this work was discovered in 1959 in the Vatican Library by the scholar Llorenç Pérez Martinez. It is part of a bunch of fascicles that are collectively known as the Codex Vaticanus Latinus 9332. It is not known when the writings came into possession of the Vatican Library, but in the fifteenth century the fascicles were owned by Pier Leoni da Spoleto, an astrologer and court physician to Lorenzo de Medici, a.k.a. Lorenzo il Magnifico.

The Codex is not easy reading. In the fifteenth century one did not buy books to build a library, one actually had to either write, or at least copy, them oneself. What is more, writing and calligraphy were a real art. Blunt pens and blotted ink made the going very difficult, and even a skilled copyist had a hard time. It probably was the medical doctor himself who transcribed—*trans-scribbled* may be a more apt word—the manuscript from the now lost original. Leoni da Spoleta's manuscript is extremely hard to study, and it took the combined efforts of philologists and handwriting specialists to decipher the obscure script and make sense of the many abbreviations. Parts of the text had to be worked out letter by letter. The tract carries no heading and its title can only be inferred by its concluding sentence: "This is the end of the method for the election of persons."

The method that Llull proposes in *Artifitium Electionis Personarum* is identical to the one presented in *Blanquerna*, except that in this text he deals explicitly with ties. This is somewhat surprising given that *Artifitium* preceded *Blanquerna*. Apparently by the time he wrote the novel,

Llull had forgotten what he had written eight to ten years earlier. Llull suggests that if the electors are evenly split in a showdown between the two candidates, each of the two receives a winning point. This means that they are judged as if both of them had won. Sounds fair enough, doesn't it? But why not award zero points to each candidate, as if both had lost? Would that not also do the trick? The answer is no, there is a subtle difference between awarding one point each or no point each. In Llull's scheme candidates who tie in one or more face-offs have an advantage over other competitors when the points are added up at the end.

This may be all right most of the time, but consider the following scenario: a mediocre candidate who ties with most competitors could receive as many points as a high-class candidate who wins most showdowns. Or consider Sister Cecilia who wins six showdowns and draws one, while Sister Dorothea draws seven times. The election would be a tie. Llull apparently thought that this is how it should be. A fairer method would have awarded half a point to each nun in case of a draw. In this case Sister Cecilia would have won with 6.5 points against Sister Dorothea's 3.5, which seems more appropriate. The idea apparently did not occur to Llull.

Now to the second sort of tie. If a standoff occurs in the final tally among two or more candidates, the latter leave the hall and the electors once more make a choice between them. (Actually the electors could have just checked the results of the initial showdowns to see who had been the winner when the two candidates competed against each other.) If again neither emerges victorious, the winner is determined through the drawing of lots. That Llull should propose that lots be drawn is somewhat astonishing, given the Christian injunction against gambling. But in this instance the devout theologian was prepared to allow divine intervention take its course in the determination of the winner . . . in the form of chance.

The election method Llull proposed may be fair but it takes an inordinately long time. Let us assume that there are twenty candidates. An exhaustive search for the most suitable candidate, as proposed in *Artifitium Electionis Personarum*, requires a series of 190 pair-wise comparisons (that is, 19 + 18 . . . . + 1). The time it takes for the two candidates to present themselves, for the electors to deliberate and make up their mind, and for the result to be entered into the tally sheet may be three to five minutes. Thus, the process may take anywhere from ten to fifteen hours.

When cutting the roster to, say nine candidates, as proposed in *Blanquerna*, only thirty-six pair-wise comparisons have to be made. This results in considerable time savings, but the election process still lasts between two and three hours.

There exists a third text by Llull that deals with elections, and the method he proposes there is much faster. The text is called *De Arte Eleccionis* (The Art of Elections) and was written in Paris, on July 1, 1299. The original of this text is also lost. Only a solitary handwritten copy, dating from the early fifteenth century, has survived. It was discovered in 1937 by Martin Honecker—the medievalist who had got Llull's translation mixed up—among the manuscripts of the library of the Sankt Nikolaus Hospital in Bernkastel-Kues, a German town known mainly for its Mosel wine. The library, one of the most valuable private collections in the world today, belonged to the Cardinal Nikolaus von Kues. It contains 314 manuscripts from the ninth to the fifteenth century. The sponsor of the library had spent the spring of 1428 in Paris, studying the works that Llull had written 130 years earlier. Maybe he even transcribed *De Arte Eleccionis* himself. I will have much more to say about the cardinal and his work in the next chapter.

Since *De Arte Eleccionis* and *Artifitium Electionis Personarum* carry similar titles, some researchers thought the two tracts were copies of the same original text. The first to actually make a comparison between the texts, using techniques from computer science, were the software engineers Dominik Haneberg and Wolfgang Reif, the library curator Günther Hägele, and the mathematicians Friedrich Pukelsheim, mentioned above, and Matthias Drton, all from the University of Augsburg in Germany. In the 1990s, they worked long and hard trying to decipher the text of Cardinal von Kues's library. In spite of their efforts, a few pieces of handwriting were either missing or illegible. Their hope was that the text from the Vatican that had been discovered in 1959 would fill the gaps. The professors asked for photocopies from the vaults of the Holy See and then waited with bated breath for the package to arrive. When it did, they were simultaneously elated and disappointed. Elated, because the Vatican text turned out to be a different text that they had never seen. Disappointed, because this meant that the gaps in the first tract would remain.

In *De Arte Eleccionis*, Llull describes what at first sight seems to be only a slight variant of the election method that he had already proposed in *Artifitium Electionis Personarum* and in *Blanquerna*. But however slight the difference may seem at the outset, it is extremely significant. While the first two texts propose a tournament in which everybody runs against everybody else, the third text puts forward a knockout procedure—pardon me, sisters—to elect the abbot, abbess, bishop, or pope. It is efficient, fast . . . and not quite fair.

The candidates enter the church, single file. The first who enters is given the letter A, the next the letter B, and so forth until the last one receives, say, the letter K. They all sit down and then the first two candidates, A and B, rise again to face each other. The electors decide which of the two they prefer. The loser is out of the race and sits down to watch the further proceedings. The candidate who wins, let's say B, moves to the next round and now runs against C. Again the loser is out and the winner moves to the next round. This continues until the winner of the next-to-last round stands against the last candidate, K. The winner of the last showdown is declared the new abbot, abbess, bishop, or pope.

The time saving is dramatic. While the method extolled in *Artifitium Electionis Personarum* required 190 comparisons and took ten to fifteen hours, and the *Blanquerna*-method required thirty-six comparisons, taking about one to two hours, *De Arte Eleccionis* entails much less effort: with twenty candidates there are only nineteen showdowns, with nine candidates only eight. The election could be over in half an hour.

The time saving comes at a cost, however. I have not mentioned it yet, but Llull's first two proposals do more than just determine the top candidate. They provide rankings of the candidates—of all of them in *Artifitium Electionis Personarum* and of the ones on the short list in *Blanquerna*. The rankings are determined by the number of wins in the two-way comparisons. The person behind the winner who garnered the next-highest number of points is declared runner-up and so on. The knockout procedure, as proposed in *De Arte Eleccionis*, only determines a winner. True, the establishment of a ranking was not a requirement of the election procedure as initially specified by Llull, but it could come in handy. If the top-ranked candidate should decide that she does not want the honor, or if she suddenly dies from a heart attack, the runner-up could simply take over. In *De Arte Eleccionis* a whole new election would be required.

Apart from the lack of a ranking, the method proposed in *De Arte Elec-*

cionis has further drawbacks. First of all, the winner may not be the same person who would have won under the method proposed in the two other texts. Only if a candidate exists who is superior to everyone else to such an extent that he would beat all other candidates in two-way showdowns, is the identical outcome guaranteed in both systems. And a real loser, an individual who would be defeated in two-way showdowns by every other candidate, won't be elected under any of the systems. That is as far as the similarities between the methods go. Apart from a very best candidate, should one exist, always winning and a very worst candidate, should one exist, always losing, the outcome is up for grabs.

And this is where a serious objection must be raised against Llull's third electoral system. The criticism is not that the winner may be different from the winner in the first two systems. The more serious shortcoming is that the election can be manipulated. In contrast to the two earlier systems, *De Arte Eleccionis* may give a second-rate candidate a good chance of emerging victorious, even if she were inferior to most of her competitors. By arranging the candidates' order of entry into the church, their chances of winning can be enhanced or diminished. Specifically, the last candidate to enter the church, *K* in the above list, must only beat the winner of the next-to-last round. If she does, she becomes abbess, no matter how unsuited she is for the job. Had she been the first to enter the church, like *A*, she would maybe have lost in the very first round to *B*. Surviving this, she could have been beaten in the second round by *C*, or the third, or the fourth. In each case she would have been knocked out. In fact, a nun could be inferior to every candidate except one, and still win the contest. And the next-to-last candidate may lose the race even if she emerged victorious from all previous showdowns.

In contrast to the method proposed in *Blanquerna* and *Artifitium Electionis Personarum*, where the order of entrance into the church does not matter, it would be advantageous according to *De Arte Eleccionis* to delay one's entrance. Imagine the melee at the church doorway. Hopefuls must arrange themselves in a single file with everybody trying to let the others pass first. The manner of influencing election results, by delaying one's entrance into the church, is an early example of what would be called "agenda setting" in the twentieth century. By deciding on the order in which candidates or options are put to a vote, one can determine, or at least influence, the outcome.

There is a way around the problem, however, surprising as it may sound. If two conditions about the electors' preferences are fulfilled, the scheme will be fair to all candidates, and the most deserving candidate will always be elected. The first condition is that there exists a true, God-given ranking among the candidates. The second is that at least half of the electors recognize this truth. If these conditions hold, the most suitable candidate will always be victorious, no matter which election method is used and in which order the candidates present themselves for the showdowns. In two-way showdowns, the majority of electors will always pick the superior candidate, no matter in which order they present themselves, such that every candidate floats to the correct position in the ranking.

The reason that problems arise if these conditions are not fulfilled is that choices among two candidates do not carry over to three or more candidates. They are not transitive. What is meant by this notion? I will explain by example. Since an elephant is heavier than a horse, and a horse weighs more than a dog, we do not have to put the dog and the elephant on a scale to check which one is heavier. Weight is a transitive characteristic of objects, and therefore we know that the pachyderm is heavier than the canine without directly comparing their weights.

Transitivity holds in many everyday situations, but not always. To illustrate intransitivity let us enter the boxing ring. If Hasim Rahman knocks out Lennox Lewis, and Lennox Lewis k.o.'s Mike Tyson, does that mean that in a direct fight Hasim Rahman would necessarily beat Mike Tyson? No, it does not. Iron Mike may have a trick up his shorts that could bring down Hasim Rahman. This is what is meant by intransitivity. Lacking the third fight, one will never know which of the two is the better boxer. The eternal appeal—or endless boredom—of the game "paper, rock, scissors" also resides in intransitivity. Paper wraps a rock, a rock scratches scissors, and scissors cut paper. None of the three items is absolutely superior to the two others, preferences cycle around and around. Sometimes intransitivity is also built into the game of poker. Usually the highest combination of cards that a player can obtain is the Royal Flush. In order to eliminate absolute certainty from the game, some people play according to the rule that even the Royal Flush can be beaten . . . by the lowest combination of cards, a single pair.

The lack of transitivity is the reason why neither of Llull's electoral methods necessarily elect the abbot and abbess who is preferred by the electorate. If transitivity holds then it is OK to use the knockout procedure for elections, because any person who wins a showdown would have also won the previous showdowns. With intransitivity the problem of cycles cannot be resolved and Llull's method does not necessarily produce an undisputed winner.

This is where things stood at the beginning of the fourteenth century. Llull's methods of an exhaustive series or a partial series of pair-wise comparisons allowed choices among more than two candidates. Thus his suggestions were a major advance over the simple majority rule or the qualified two-thirds majority rule that had been in use since ancient times. But they had their shortcomings. The first method, advocated in *Blanquerna* and in *Artifitium Electionis Personarum* is intransitive, which means that it does not resolve the problem of cycles. The second method, described in *De Arte Eleccionis* is, in addition, not fair to those who get to church early.

BIOGRAPHICAL APPENDIX

Ramon Llull

As a young man, Llull had a reputation for high living and made a name for himself as a troubadour "composing worthless songs and poems and doing other licentious things," as he would later say in his autobiography. Even after he was already married to one Blanca Picany with whom he had two children, Dominic and Magdalen, he pursued ladies of the court with undiminished attention. But one day— he was thirty years old at the time— a vision changed his life forever: in an apparition he saw Christ on the Cross. Llull decided to forget about chasing women and devote his life to three aims: firstly, "to accept dying for Christ in converting the unbelievers to His service"; secondly, "to write a book, the best in the world, against unbelievers"; and thirdly, "to procure

the establishment of monasteries, where various languages could be learned." The reason for his third aim will become clear presently. Having decided on this career change, he sold his entire belongings, except for a small portion that would serve to support his family, said goodbye to his wife and children, left the merry life at court, and henceforth pursued a monastic lifestyle.

Converting Muslims and Jews to Christianity became an obsession to which Llull devoted all his efforts. He decided to follow a novel and, for his times unique, approach. Instead of crudely pointing out to the infidels that Christian doctrines were superior to the tenets of their own faith— arguments usually emphasized by fire and sword—Llull would attempt to make them see the light by engaging them in discussions and rational arguments. He was of the opinion that even though bloody crusades may be indispensable, they should be accompanied by attempts at persuasion. But before he could actually apply this subtle approach to proselytization, he made an important discovery: in order to convert Muslims he would have to learn Arabic. This is the reason why he aimed to create language schools.

He immediately set out to learn Arabic, and a Muslim slave was employed to give him lessons. But his initial language classes were met with a mishap. The sessions were not limited to learning grammar and syntax; instead, they branched out to questions of faith. On one occasion, the slave uttered a blasphemy. An argument ensued that got so heated Llull slapped him, whereupon the slave procured a sword and tried to kill his master. But Llull was no faint-hearted monk. Even though he was seriously wounded in the stomach, he managed to subdue the aggressive slave and, together with servants who had come running, lock him up. While he was still contemplating what to do next, disaster struck: the prisoner hung himself in his cell. It was both a tragedy and a relief for Llull. The tragedy was that he had not been able to win the infidel over before his death. The relief was that the decision of what to do with the blasphemous slave had been taken out of his hands.

Llull gave thanks to God and henceforth lobbied for the creation of language schools, so that missionaries could learn Arabic in a calmer

setting before they would set out on their missions. At his urging, King James I of Aragon founded a school for the study of Oriental languages in Llull's homeland of Majorca. The first class consisted of thirteen Franciscans who also studied the liberal arts, theology, and Islamic doctrines. They also undertook an in-depth study of Llull's very own *Ars inveniendi veritas* (Method of finding truth). Some thirty years later, the council of Vienne (1311–12) would establish chairs for Oriental languages at the universities of Paris, Bologna, Oxford, and Salamanca.

Llull's missionary tactic was not to refute the doctrines of "unbelievers" by basing his arguments on the sacred texts *he* believed in. He realized that attempts to convert Jews and Moslems by pointing out the purported superiority of Christian texts and by rubbing their noses in the alleged errors of the Talmud and the Koran had generally been unsuccessful. Rather, he sought the common ground. After all, there was a lot the three religions could agree upon: they believed in only one God, they had similar notions of what constitutes virtues and vices, and everybody knew, of course, that the Earth was the center of the universe. Llull was convinced that the three monotheistic religions worshiped an identical God and that only the forms of worship differed. Today Llull is considered one of the most tolerant theologians of the Middle Ages. He was not prepared to grant other religions a right to exist, mind you, that would have been too much to ask for from a medieval theologian, but he did admit that there existed learned Jews and

Muslims. If they would only listen to him, he thought, they would soon come to realize that Christian doctrines were simply superior to anything they believed in. His problem was that not everybody was prepared to listen. But at a time when the preferred method of spreading the faith was to beat eternal truth into nonbelievers, Llull's soft style, combining theological, scientific, and moral lines of reasoning, was quite a novel approach.

Before he could start teaching Christianity, however, Llull needed to acquire academic credentials. Thus he undertook studies at the universities of Montpellier and Paris and was thenceforth known as a *magister*. Now he was ready to evangelize the Saracens, as Muslims were called during the Middle Ages. He arrived in the port city of Genoa with the intention of setting sail for Muslim lands where he would put his new skills to use. The town was abuzz with news of his arrival. People were duly awed by this learned man who was going to convert the unbelievers to the faith of Christ. But then things turned out a little differently than planned. At the first chance Llull got to travel, he reneged. With his belongings and books already on board a ship, he suddenly became afraid that the Saracens would slaughter him the moment he arrived, or at the very least throw him into prison forever. Forgetting his previously mentioned intention to die for Christ in converting the unbelievers, he stayed put in the port city of Genoa.

The scandal that ensued threw Llull into a deep depression. He became gravely ill for a long time. Eventually he mustered his courage and traveled to Tunis, where he finally managed to get involved in debates with Islamic scholars. They were not successful. Maybe he was not yet quite fluent in Arabic, or maybe the scholars did not find his arguments very convincing. In any case, Llull was soon banished from Tunis and told, under penalty of death, never to come back. He nonetheless secretly returned, hoping to complete his mission by baptizing several men of considerable reputation who had already agreed to accept the Christian faith. But after three futile weeks, during which another Christian man, resembling him in dress and demeanor, was nearly stoned to death, Llull left for Naples. He tried his luck in Cyprus next, but was also not very successful. After an attempt on his life he left the island and continued on. By some accounts he traveled as far as Jerusalem before returning to Europe.

Llull experienced many adventures, was imprisoned at various times, shipwrecked, or otherwise laid up. His arguments about the relative merits of the different faiths did not gain in popularity even as time went by. But he never wavered. "I was married and with children, reasonably well off, licentious and worldly," he wrote in one of his last works. "All of this I willingly left in order to honor God, procure the public good, and exalt the Holy Faith. I learned Arabic, several times I ventured forth to preach to the Saracens; and for the sake of the Faith I was arrested, imprisoned, and beaten. For forty-five years I have labored to move the Church and the Christian princes to act for the public good. Now I am old and poor, yet my purpose is still the same, and the same it will remain, if it so please God, until I die."

With advancing age, the *Doctor Illuminatus*, as he was now known, became less tolerant. The futility of his efforts to proselytize through rational argument made him seek a more effective method. He finally started advocating crusades as an appropriate technique to pull Moslems over. Thus most of his later life was spent trying—unsuccessfully—to convince the powers that be to put together armies for that holy purpose. But he never abandoned his penchant for the soft approach, always pointing out that violence should be used only if absolutely necessary.

His participation in discussions was never met with much enthusiasm by his opponents. At one disputation in 1315—once again in Tunis—the unfortunate missionary, who was more than eighty years old by then, was stoned by the Moslems. Left for dead, the badly wounded man was found by Italian merchants who transported him to their ship. The vessel set sail for Majorca but Llull was only able to catch a glimpse of Palma before he passed away. Actually, this last chapter of his life may not be quite true. It was probably embellished by his followers to aid them in their quest to have the Catholic Church canonize Llull.

The researchers at the University of Augsburg prepared a beautiful Web site in which the three texts—*Artifitium Electionis Personarum*, *Blanquerna*, and *De Arte Eleccionis*—can be viewed simultaneously in the original handwriting, transcribed into proper Latin, and translated into English, French, or German. Whenever the user clicks on a sentence or word in any of the three windows, the appropriate portions in the other two windows are highlighted (www.uni-augsburg.de/llull).

CHAPTER FOUR
THE CARDINAL

After Llull's foray, no further progress was made in the theory of voting and elections for over a century. Then, in 1428, the German student Nikolaus Cusanus happened upon one of Llull's documents in Paris. He found the text sufficiently interesting to make a transcript and take it home. But he did more than reproduce the text for his own use. A few years later he improved upon Llull's method, thereby establishing himself as the second pioneer of the modern theory of elections.

The young man was born as Nikolaus Krebs in the German town of Kues in 1401, 102 years after Llull wrote *De Arte Eleccionis*. The original name in German means crab, which is why a crustacean is featured on the family's coat of arms. Krebs was never ashamed of his petit bourgeois background, but as he got older he found it appropriate to adopt a nobler surname. He started calling himself Nikolaus von Kues, or Cusanus, which had a better ring to it than the original.

His father was an upwardly mobile businessman who, by some accounts, was rather strict. Nikolaus entered the University of Heidelberg at age fifteen to study the arts. A year later he went to Padua, Italy, to get a law degree, and then continued on to Cologne where he studied philosophy and theology. At all the institutions he attended, he made intensive use of the archives, becoming a very learned man in the process.

As a youth, Kues had witnessed one of the Catholic Church's more embarrassing moments, the Papal Schism. No less than three men claimed to be pope. It was only through convoluted maneuvers that the delegates to the Council of Constance (1414 to 1418) managed to resolve the issue by kicking the three reigning Holy Fathers—Gregory XII, Benedict XIII, and John XXIII—from their Holy Sees. But deposing the feuding popes proved to be the easy part. In order to install a new church leader the council had to take recourse to creative, and controversial, election methods. Since Italian churchmen had traveled to Constance in great numbers in order to cheer along their preferred pope, John XXIII, the other nations' delegates felt the need to devise clever ways to contain the Italians' electoral power.

The French argued for a widening of the electoral body by including doctors of theology and all priests present. That proposal did not meet with the approval of the Italians and the Germans and the feuds continued. Finally it was resolved that the vote would take place by nations. The twenty-three, mostly Italian, cardinals were augmented by five deputies from each of the six nations present.

On November 8 *anno domini* 1417, the fifty-three electors retired to a conclave, a closed hall especially prepared for the election. It took them three full days to agree on a personality from among their midst, the Roman Cardinal Odo Colonna, who became Pope Martin V. Actually three days, long as it may seem, is not too bad when compared to the two years and nine months it had taken to elect Gregory X in 1271. (And even then the quarrelling cardinals from France and Italy could be persuaded to settle on a candidate only after the town folk of Viterbo, where the election took place, put them on a diet that consisted exclusively of bread and water.) This experience may have planted in Cusanus the desire to seek a more efficient manner of electing church personnel and other officials.

It was at the next congregation of churchmen, the Council of Basel, convoked by Pope Martin V in 1431, that Cusanus first made a mark. As long and drawn-out as the four-year period in Constance may have been, it was nothing compared to what would happen in Basel. The council lasted for a full eighteen years. Cusanus went to Basel in an attempt to settle a dispute about the election of the bishop of the town of Trier. The loser of that election, sorely disappointed at his defeat, had hired Cusanus as a lawyer to argue his case. In the early fifteenth century it had not yet been decided whether the pope or the council should have the last word in questions of faith and administration of the church, so Cusanus had to choose to whom to turn. After due consideration he sided with the council and instead of traveling to Rome to argue his client's case, he journeyed to Basel. It was the wrong choice and the proceedings ended in defeat. However, in one respect at least, the lawsuit did produce a positive outcome for Cusanus, if not for his client. As we shall see, he managed to establish himself as a scholar of note.

Cusanus was deeply upset about the loss of his case but did not resign himself to his fate. In the course of the proceedings he had taken a liking to the churchmen and decided then and there to abandon the practice of law in order to follow an ecclesiastical career. He continued to support the council at first but, over time, became disenchanted with its attempts to ostracize the papists. Such efforts ran counter to his belief that the church needed to be unified. So, five years after throwing his lot in with the council, Cusanus changed his mind and became a fervent papist. His support of the pope did not derive from a conviction that the Holy Father was infallible. Rather, he believed that the unity of the church could best be served by entering the services of the Vatican.

The Holy Father assigned several missions in Germany to Cusanus. His task was the reform of the church, the orders, and the convents. He became a workaholic, and his zeal and relentless efforts were not universally appreciated. The locals took to calling him the Pope's Hercules against the Germans. But in Rome at least, his work was valued highly. In recognition of his achievements the pope named Cusanus cardinal in 1448. All of a sudden his compatriots grew more proud of him. They now fondly called him the German Cardinal (because most other cardinals were Italians, of course). As a senior dignitary of the church, Cusanus railed relentlessly against the churchmen's greed and the accumulation of sinecures. Eventually, he came up with a rather revolutionary idea: church officials should perform all services free of charge. If funds were needed, they should be raised through voluntary contributions. This sounded politically very correct but then something unexpected happened. With the rise to power, Cusanus started to develop a taste for worldly riches.

Cusanus was called back to Rome. His benefactor, Pope Pius II, dreamed of a crusade against the Turks. The cardinal, who leaned toward tolerance of Jews and Muslims, quite in the spirit of Ramon Llull, hated the idea. But his vows bound him to the pope and, following the orders of the Holy Father, he left Rome to gather warriors under the church's command. On the way to Ancona, Italy, where the crusader army was to meet up with the Venetian navy, Cusanus became violently ill. On August 11, 1464, he died in the Umbrian town of Todi. All his belongings, among them his valuable library, were bequeathed to the hospice for the poor in his hometown Kues. During World War II, the allies, who were aware of the treasures hidden away in the town, abandoned plans to bomb the area. This is how Cusanus's manuscript survived until today.

In Basel, Cusanus wrote his major work, *De Concordantia Catholica* (On Catholic harmony), with which he established himself as a true scholar. His aim in writing the book was the unity of church and state. As I describe in the additional reading section of this chapter, the part of *De Concordantia Catholica* that eventually made Cusanus famous could have mightily displeased the powers that be, and he was extremely fortunate that it turned out otherwise.

One of the book's themes was elections. His interest in the subject had been kindled in 1428, while still a student, when one of his teachers sent the promising young man to Paris for further studies. Cusanus duly traveled to the French capital and made good use of all the city had to offer. He toured the libraries and in an archive he happened upon one of Llull's manuscripts. The writings of the Catalan mystic had always held a certain fascination for Cusanus, but what he found in the library in Paris had a most profound influence on him. So impressed was he by Llull's *De Arte Eleccionis* that he decided to take a copy back home. Had it not been for this, we would never have known about Llull's text, because Cusanus's transcription is the only copy that is known to exist.

Five years later, when working on *De Concordantia Catholica* in Basel, Cusanus grappled with the question of fair elections again. At first he pondered Llull's method of pair-wise showdowns, but then decided on a different scheme. A sizeable part of *De Concordantia Catholica*, chapters 36 and 37, are devoted to the subject. The method that Cusanus came up with was framed not as the choice of an abbot, bishop, or pope, but as the election of a monarch, the Holy Roman Emperor.

Let us suppose, as Cusanus did, that ten candidates aspire to the job of emperor. An unspecified number of electors, the German Kurfürsten, meet in a hall to choose the one who would be the most worthy from among the ten. Every elector receives a set of ten slips of paper, each of which carries the name of one of the candidates. With the slips of paper under their arms, the men wander off into the corners of the room or into a corridor in order to make up their minds. Each elector ponders who in his opinion is the least suitable candidate, takes the slip carrying that person's name and puts the number one onto it. Then he decides who is the next-least worthy candidate and writes the number two onto the slip that carries his name. In this manner he continues until there is only one slip left. Obviously it carries the name of his preferred candidate. He writes the number ten onto it, goes back to the voting hall, and puts all slips into a bag that hangs in the middle of the room.

Once the electors have deposited their slips, a priest, beyond any suspicion of dishonesty by virtue of his office, empties the bag and announces the names and numbers on each slip. An assistant conscientiously jots down the announcements. After all slips have been registered, the points are added up and the candidate who amassed the most is declared emperor. In effect, what the electors did was to award a number of points to the candidates, commensurate to how they evaluate their worth. The better the candidate in the elector's judgment, the more points he gets. This is the crucial difference to Llull's scheme where the winner of every showdown receives just one point, no matter how superior he or she is. In Cusanus's scheme, the difference in points awarded to two candidates by an elector is determined by how far apart they are on his ranking.

In order to avoid *practicas absurdissimas et inhonestissimas* (most absurd and dishonest practices) Cusanus instructed the electors to make their choices in complete secrecy. He also recommended that the electors write with similar pens, and use similar strokes to write the numbers onto the slips, so no-one's writing could be recognized. Thus secrecy was assured. Exceptions would be permissible only for illiterate electors. They were to be allowed the company of a trusted secretary who would read to them the names on the slips. Cusanus wrote this without condescension, by the way, even though there was the saying in his times "*Rex illiteratus est quasi asinus coronatus,*" which means, in English, an illiterate king is like a crowned ass. But in the high Middle Ages, power in Germany did not necessarily go hand in hand with a good education.

Why does secrecy guarantee fairness? After all, Llull also desired fairness but recommended the exact opposite: open elections. Who is right? The short answer is: they both are. And now to the long answer. Cusanus justified secrecy with arguments that also hold today when a group of people who do not trust each other elect an official. If votes are cast in secret, electors cannot sell their votes and cannot be threatened. Actually, they could offer to sell their votes but there would be no takers since compliance with some deal or arrangement could not be verified. Also, threats *could* be made, but they would not scare anybody, since electors, hidden away in the corner of an election hall or in a booth, can mark the ballots any way they want. (This is why secrecy in an election is not just a right but also a requirement. It is illegal to show your ballot even if you agree to do so, because then you could prove your compliance to some deal.)

Llull had a different scenario in mind when he designed his procedure.

While Cusanus's method was meant for distrustful electors who only met once every ten or twenty years in order to elect a king or an emperor, and would part ways again after the election, Llull's scheme was intended for monks, nuns, and friars. The members of these societies would have to continue to live and work together after the election. One requirement of a smooth organization is that the members trust each other. And what better way to seam discord and mistrust than to promise one's vote to numerous candidates and then secretly cast a different vote altogether. With open elections, nobody will dare cast a vote with ulterior motives in mind. Everyone would immediately know, and the elector's popularity rating would fall accordingly. This is the reason why directors on company boards or members of committees nowadays usually vote in full view of their colleagues. Secret back stabbings are not possible and the electors are held accountable for their votes. Thus they have faith in each other's judgments and can continue working together after the votes. (This may be a somewhat idealistic view of things. To know that one's closest friend voted for someone else may not be conducive to trust and harmony.)

Now back to Cusanus. In order to carry out their rankings, the electors start by spreading out the slips with the candidates' names in front of them. Let's say they place them in a vertical column. At the outset the slips are arranged in a completely random fashion. Then the work starts. A good, if not very efficient, method of ranking candidates is the following: starting at the top, compare each pair of successive candidates. If the slip carrying the name of the better candidate lies below the other slip, exchange the slips carrying their names. Otherwise leave the two slips in the order they were in. Move down the column in this manner until you reach the bottom. Then start at the top again. The procedure must be performed over and over, until a run through the whole list has been completed without a single inversion. At that point the elector is satisfied with the ranking and his work is completed.

This method of arranging candidates into an order with the best on top and the worst on the bottom is what is called the "bubble sort" in computer science. The bubble sort is an algorithm, a recipe that arranges a list of items in ascending or descending order. It derives its name from the fact that during the runs the best item—in our case the slip with the preferred candidate—eventually rises to the top like a bubble in a liquid. Unfortunately this may take a very long time. In fact, the bubble sort is very,

very slow. Budding computer scientists are warned never to make use of this inefficient algorithm, except to show how inefficient it is. There are many sorting methods known to computer scientists that are much faster: the insert sort, for example, or the shell sort, the heap sort, the merge sort, the quick sort. Had they been known at the time, any one of the algorithms could have been used by the electors. But even with the more efficient methods, ordering the slips could take quite a while if the number of candidates is large.

There is a redeeming feature to Cusanus's scheme, however: the electors make their rankings simultaneously. Nobody wastes any time, waiting until his predecessor is done. They all finish more or less at the same time, and then all that needs to be done is to aggregate the results. If the electors need about twenty minutes to rank ten candidates, and the priest takes another half an hour to aggregate the results, the election could be over in less than an hour.

Since this scheme does not involve two-by-two showdowns but overall evaluations by the electors, the Cusanus-winner may be different from the winner in either of the two Llull schemes. But the method does more than designate a winner. As was the case with the first two of Llull's schemes, it has the advantage of providing a ranking of the candidates according to the number of points the candidates rake in.

Cusanus ignores two potential pitfalls. One may occur if an elector considered two candidates equally worthy, thus not wanting to place either above the other. The other could come about if an elector preferred, say, Rüdiger to Sigismund, Sigismund to Bernhard, and Bernhard to Rüdiger. If that were to happen, the poor elector would constantly shuffle around his slips of paper without ever arriving at a satisfying ranking. Cusanus simply assumes that such situations do not arise.

There is another implicit assumption in Cusanus's scheme, namely that each additional rank merits exactly one additional point. Other methods of assigning points to ranks are conceivable. Cusanus could have suggested, for example, that the three candidates ranked at the bottom receive no points at all, or that the top-ranked candidate receive an additional two points, or three points, or whatever. Infinitely many possibilities exist to award points, of which Cusanus chose the simplest. For example, the deadly boring Eurovision Song Contest on European television uses a variant of Cusanus's method to assign points. More than three dozen countries participate in the contest by sending singers to the venue and assigning a jury to judge the songs. In a first round, all but twenty songs are eliminated. In the second round, a week later, the juries are not allowed to vote for their own country's entry, so each chooses among nineteen songs. The eight songs, that are worst in a jury's opinion—and believe me, they are dreadful—are given no points at all. Then each jury deals with the eleven songs that it has decided are not quite as bad as the really dreadful ones. While 200 million viewers throughout Europe wait with bated breath, each jury awards one point to the song it ranks lowest, two points to the next lowest, and so on all the way to the second best song, which receives ten points. Finally, the best song receives not eleven, but *twelve* points. Once the twenty juries have announced their rankings, the points are added up and the winner is declared. This is the scheme proposed by the cardinal, except for the dreadful songs, which receive zero points, and for the rankings in first place, which receive one more point than they should have according to Cusanus. Apparently the organizers of the Eurovision contest felt that making it to "best song" in any jury's opinion is worth an additional bonus point.

Obviously, the manner in which additional points are awarded to a better rank can change the outcome of the contest during the aggregation process. In the Eurovision contest, for example, a song that ranks 1st and 11th, respectively, in two juries' opinions, obtains thirteen points, while a song that ranks 2nd and 10th place receives only twelve. In Cusanus's original scheme, the two songs would have drawn with twelve points each. So not only can the Cusanus-winner be different from the Llull-winners, but in different variants of Cusanus's scheme different winners are possible.

There remains one issue that has to be resolved. As we saw at the end of the previous chapter, the German medievalist Martin Honecker believed, incorrectly, that in the scheme described in the novel *Blanquerna* Llull had suggested that individual votes were to be counted, not just the wins. This is exactly what Cusanus proposed a century later. So did he simply copy Llull, albeit misreading and mistranscribing the text, as Honecker would do 600 years later? Did Cusanus also understand *veus* to mean "votes of the electors" instead of "wins"? If this were the case, then Llull would have been a closer precursor to Cusanus than is generally believed, and Cusanus would have been no more than a plagiarizer. The English scholars Iain McLean and John London believe that this is not to the case. In a paper that appeared in 1990 in the journal *Social Choice and Welfare* they argued that Cusanus was probably aware of the existence of the novel *Blanquerna* because it appears in a listing in his library. But they are convinced that he had not read, let alone misread, it. Hence, he had come up with his scheme independently.

Let me recall the main points of this and the last chapter. In *Artifitium Electionis personarum* and in the novel *Blanquerna* Llull proposed exhaustive series of pair-wise comparisons among contestants. The candidate with the most wins is elected. In *De Arte Eleccionis*, Llull suggested a knockout procedure. A century later, Cusanus proposed a method in which each elector first ranks the candidates, and then awards them a number of points according to their ranks. The candidate who garners the most points is elected. As I pointed out, both of Llull's methods suffer from intransitivity, while Cusanus's scheme is lacking in that it arbitrarily assigns one additional point for every better rank. It is not surprising that the three methods may result in different winners.

With the cardinal's passing, the theory of voting and elections went into hibernation for a few hundred years. Buried in manuscripts that nobody bothered to read, the contributions of Llull and Cusanus were completely forgotten. The next development occurred only toward the end of the eighteenth century, in France.

ADDITIONAL READING

De Concordantia Catholica

The part of *De Concordantia Catholica* that eventually made Cusanus famous could have mightily displeased the powers that be. The then thirty-year-old gave the first proof of his erudition by debunking a forgery of the church, the *Donation of Constantine*. This document, supposedly written in the fourth century AD, alleged that King Constantine (ca. 274–337) made a substantial and very important gift to the church. Before he moved east to found the city of Constantinople (today's Istanbul) Constantine suffered from leprosy. In his despair, so the story went, the unbelieving king turned to Pope Sylvester. The Holy Father, not at all versed in medicine but an authority on all matters of faith, convinced him that the best treatment for his ailment would be conversion to Christianity. Constantine had himself duly baptized and was promptly healed from his ills. Now comes the clincher: out of gratitude, he donated his palaces, the city of Rome, and the western part of the Roman Empire to Pope Sylvester. He had real estate to spare, after all, and should he require additional territories, there would always be other lands to conquer.

For the church the *Donation* was a gift sent from Heaven. Without any benefactors to speak of, there was nobody to fight its wars and the pope had to look out for himself. In such situations the document provided some sort of legitimacy for the church's territorial claims. The *Donation of Constantine* also meant that Charlemagne (742–814), or "Karl der Grosse" as he was known in Germany, the founder of the Holy Roman Empire of the German Nation, drew his legitimacy from the pope who had crowned him Emperor in 800. All through the Middle Ages, whenever anyone dared doubt the church's supremacy over the worldly power, one church man or another simply whipped out the document and all questions were silenced.

The deed served the church well until 1433 when our young hero from Germany decided to subject the document to closer scrutiny. Studying not only the text itself but also all relevant contemporaneous literature, he

made a momentous discovery. Based on comparisons with facts that would have been unknown in the fourth century, and information that only a later scribe could have been aware of, Cusanus concluded that the *Donation* was a forgery written in the eighth century. Actually, the story about Constantine and Sylvester should have raised an eyebrow or two much earlier since anybody who had inspected the sources could have known that the emperor neither suffered from leprosy nor became a Christian during his lifetime. Constantine was converted to Christianity only on his deathbed. But apparently nobody really wanted to know the truth.

The momentous discovery that the church had not hesitated to stoop to forging documents in order to further its interests did not have the negative impact on Cusanus's career that one would have expected. Only seven years later, when the scholar Lorenzo Valle published similar findings—albeit in a more combative and polished style—did the church start to take notice. But even then the truth was suppressed for another few hundred years, and at least one person was burned at the stake for daring to raise questions about the document's authenticity. Fortunately Cusanus was spared any ill consequences. In fact, the exposure of the church's fabrication catapulted him to prominence.

THE OFFICER

The eighteenth century was a period of enlightenment throughout the Old and New World. France, the United States, and Poland granted themselves constitutions. Nations were in upheaval as their citizens started demanding equal justice for all, showing concern for human rights, and calling for a regulation of the social order. At the same time, demands for quality government arose and the question of how officials were to be elected to high positions became important again. In this atmosphere two eminent French thinkers appeared on the scene. One was a military officer with numerous distinctions in land and sea battles. His name was Chevalier Jean-Charles de Borda. The other was the nobleman Marquis de Condorcet. The two men, outstanding scientists in Paris during the time of the French Revolution, did something amazing: they reinvented the election methods that Llull and Cusanus had proposed a few hundred years earlier. Actually, they did more than that: they provided the appropriate mathematical underpinnings. At odds with each other on many subjects, they also engaged in a lively debate on the theory of voting and elections.

Born in 1733, Jean-Charles de Borda was the tenth of sixteen children. His parents, both of whose families belonged to the French nobility, were Jean-Antoine de Borda, Seigneur de Labatut, and Marie-Thérèse de la Croix. The boy exhibited great enthusiasm for mathematics and science at an early age and a cousin of his, Jacques-François de Borda, who was in touch with the leading mathematicians of his time, was to point Jean-Charles in the direction of his future career. Jacques-François tutored the boy until, at age seven, he was ready to enter the school of the Barnabite Fathers, whose curriculum was for the most part limited to the teaching of Greek and Latin. Four years later, it was Jacques-François again who convinced Jean-Charles's father to send his son to the Jesuit college of La Flèche, where the offspring of noblemen were educated. It was there, finally, Jean-Charles received a solid grounding in mathematics and the sciences. His achievements were far above average and upon graduation the

Jesuit teachers encouraged the fifteen-year-old boy to enter their order. But Jean-Charles had no interest in religion. He wanted to continue the family tradition and enter the military. The French army provided career opportunities not only for brave fighters but also for intellectuals. Jean-Charles's father allowed his son to follow his own wishes even though he had wanted him to become a magistrate. Thus began Borda's career as an army mathematician.

When Borda was twenty, his first mathematical paper, a piece on geometry, came to the attention of Jean le Rond d'Alembert, the renowned scientist in Paris. Three years later, while serving in the cavalry studying the flight path of artillery shells, Borda presented a theory of projectiles to the Académie des Sciences, whose members elected him to its ranks on the basis of this work.

But his calling still was the army, and the young officer climbed the rungs of the military's hierarchical ladder. As aide-de-camp to the Maréchal de Maillebois, Borda participated in the battle of Hastenbeck in July 1757, where the French army defeated the Duke of Cumberland. But by then he had had enough of horses and decided to exchange the cavalry for the sea. Completing the navy's two-year course in one year he devoted himself to naval construction and to the study of fluids. The navy was suspicious of this "terrestrian" who sought to gain entry into its close-knit officers' corps. Borda managed to prove himself through his academic achievements, however. Taking issue with Newton's theory of fluids for example, he proved that a spherical body offers only half the resistance to airflow than a cylindrical object of the same diameter, and that the resistance increases with the square of the velocity. By advocating spherical shapes, Borda became an early pioneer of submarine and airplane construction. Bodies of this shape would dominate travel under water and in the air—at least until it was discovered that for supersonic flight the most efficient aircraft body has a pointed shape.

The young officer also dealt with more prosaic gadgets like pumps and waterwheels. Throughout his life, Borda participated in many voyages, battles, adventures, and scientific endeavors. For now, we limit the narrative of his achievements to his preoccupation with elections, postponing other parts of his colorful life to the chapter's additional reading section.

The French Revolution, which cost so many of his contemporaries among the nobility, the officer class, and the scientific establishment their lives, left Borda unscathed. He did not participate in any political activity, sitting out the eleven months of the great Terror (September 1793 to July 1794) in his family estate in Dax, a town in the southwest of France. After a long illness, he died in 1799. Many internationally known scientists attended the funeral below Montmartre. His scientific achievements include important advances in experimental physics and engineering, in geodesy, cartography, and other areas.

Staying aloof of political matters during the revolution did not mean that Borda was uninterested in the political process. In fact, it was a sign of the times that one of the areas he dealt with was the theory of voting. In 1770 he had already delivered a lecture about his ideas on a fair election method before the Academy of Sciences. Too busy with military matters at the time, he neglected to publish anything, however. It was only eleven years later, in 1781, that Borda wrote an article titled "Mémoire sur les élections au scrutin" (Essay on ballot elections), which was published in the *Histoire de l'académie royale des sciences* three years later. A preface to Borda's paper, written by an unnamed discussant, lauded it profusely. The introductory essay ended with the sentence that Monsieur de Borda's observations about the inconveniences of election methods that had been nearly universally adopted are very interesting and absolutely new. (The discussant is nowadays believed to be the Marquis de Condorcet, hero of our next chapter.)

In the paper Borda analyzed the well-established method of electing a candidate to a post by majority decision. It seemed obvious to most that this was the correct and fair manner to elect officials. But was it really? Should majority decisions be accepted without question? Borda took issue with the basic, universally accepted axiom that underlies ballot elections, namely that the majority of votes expresses the wish of the electorate.

The axiom seemed reasonable enough and nobody ever made any objection to it. Everybody was convinced that the candidate who obtains the most votes is necessarily preferred to all competitors. It came as a great surprise, therefore, when Borda showed that very often this is not the case. In fact, he maintained that the method of majority elections is unquestionably correct only if no more than two contenders run for a position. If three or more people present their candidacies, Borda pointed out, majority decisions may lead to erroneous results. To make his point, he presented an example in which a paradoxical situation arises. It is not at all contrived and can easily appear in everyday elections.

I will illustrate Borda's example with the election for class president at a high school. The class comprises twenty-four students. Peter, Paul, and Mary vie for the post; the twenty-one remaining students have to decide among them. Of course they use the age-old method of majority voting. Everyone puts the name of the preferred candidate on a piece of paper and drops the ballot into an urn. The count reveals that eight students voted for Peter, seven for Paul, and the remaining six for Mary. Peter, smiling broadly, thanks the voters for their confidence while Mary, disappointed at her poor showing, leaves the classroom in tears. But is the will of the twenty-one electors truly reflected in this result?

Let us poll the students more deeply about their preferences among all three candidates. The following becomes apparent. The eight students who put Peter first, would have put Mary second and Paul last. The seven who voted for Paul would also have put Mary second and relegated Peter to the end of the list. Finally, Mary's six supporters would have placed Peter behind Paul. The complete list of preferences can be summarized in the following table (where "preferred to" is indicated by " > "):

8 electors:	Peter > Mary > Paul
7 electors:	Paul > Mary > Peter
6 electors:	Mary > Paul > Peter

If we now scrutinize the preferences, we realize that in direct showdowns as advocated by Ramon Llull (see chapter 3), both Mary and Paul would have beaten Peter by thirteen votes against eight. (The seven electors in the second line of the table, and the six electors in the third line, place Peter behind both Paul and Mary.) So Peter, the undisputed winner of the majority vote, would already be out of the race. A comparison of the voters' preferences between Mary and Paul would then reveal that fourteen classmates (those in the first line and those in the third line) would have voted for her, and only seven for Paul. Now the shoe is definitely on the other foot: Mary wins, and Peter surreptitiously wipes away a tear. The results are the exact reverse of the ones obtained by majority election.

The simple explanation for this paradoxical situation is that the support Peter receives from eight electors is more than offset by the utter rejection of his candidacy by thirteen others who place him dead last. The paradox had gone unnoticed for centuries because once an election was over, nobody ever bothered to compare the voters' preferences among the losers. There will be more to say about this paradox in the next chapter.

In one fell swoop Borda challenged an election method that had been used throughout the world for centuries. The example shows that different outcomes may occur as soon as voters become more farsighted and take into account the preferences beyond their first choice. Borda compared the situation to a sporting event in which three athletes vie for the title. After two competitors have worn themselves out in a first bout, both of them, by then tired and exhausted, may succumb to a weaker opponent.

But the navy officer did not simply criticize the age-old method, he also proposed a remedy. He called it *"Éléction par ordre de mérite"* (Election by ranking of merit). It would lead to a great debate between two outstanding French intellectuals of the eighteenth century.

In Borda's proposed voting method every elector jots down the names of the candidates in the order of merit he accords them. The ranking could be, for example, Peter on top, then Paul, then Mary. Borda proposed awarding one merit-unit—let's call it an *m-unit* for short—to each rank. Mary at the low end would get one, Paul two, and Peter three. If more candidates are present, the count would go higher. For eight candidates, the bottom-ranked candidate receives one m-unit, the top-ranked candidate eight.

This manner of awarding m-units rests on an assumption, however. Borda maintains that the degree of superiority the elector accords Peter over Paul is the same as the superiority of Paul over Mary. This assumption needs a justification and Borda provides it by employing some hand waving: since there is no reason to rank Paul (the middle candidate) closer to Peter than to Mary, the correct method would be to place him smack in the middle between them. So, given Borda's belief that the difference in merit between all ranks is identical, it is quite reasonable to award one additional m-unit to the next-higher rank. Of course, many people would dispute this assumption. The *intensities* with which electors prefer one candidate over another may differ.

Now on to the next stage. Peter in the above example was ranked first by eight, and last by thirteen electors. In Borda's manner of reckoning

Peter would therefore obtain 37 m-units ($[8 \times 3] + [13 \times 1]$). Paul would receive 41 ($[8 \times 1] + [7 \times 3] + [6 \times 2]$) and Mary a whopping 48 ($[8 \times 2] + [7 \times 2] + [6 \times 3]$). Now we understand why Mary should win. By the way, this manner of adding m-units also rests on an assumption: electors are considered to be equal. Then m-units awarded by different electors have the same value and can be summed. (Many people would dispute this assumption also. After all, my m-units may be different from yours.)

The astute reader may have recognized in Borda's count of m-units the method from the previous chapter, put forth by Cardinal Cusanus. The French navy officer was not aware of his predecessor. Indeed, the Cardinal's proposals for the election of popes and emperors were quite unknown during Borda's times and only rediscovered in the late twentieth century. But Borda would have provided an advance even if the earlier writings had been known to him. While Cusanus had implicitly assumed that one additional rank—be it from rank fifteen to fourteen, or from rank two to one—should always accord the candidate the same additional gain, Borda made the assumption explicit and gave it a justification. Granted, it was a hand-waving justification, but a justification it was nonetheless. The method suggested by the cardinal and by the navy officer, of choosing among candidates by assigning points, or m-units, according to their standing in the electors' rankings, is nowadays known as the Borda count.

Borda's and Cusanus's assumption that each additional rank is worth the same is crucial. Without it, several variations of the method can be thought of. The Eurovision song contest, mentioned in chapter 4, is a case in point. There, no m-units are awarded to the worst songs. Then one m-unit is given to the song ranked eleventh, and one additional m-unit is awarded for each rank up to the second-best song, which receives ten. Finally, the best song in a jury's opinion receives twelve m-units. Different variants of the method could be thought of, and they may result in different winners.

After presenting his method of voting "by order of merit," Borda went on to analyze under what circumstances the winner according to his scheme would coincide with the winner in a majority election. How many votes would a candidate need to receive in a conventional majority election so that he would also be guaranteed the top spot according to the rules of the Borda count? Let us say that Peter and Mary are ranked in first place by a and b electors, respectively. Borda investigates the worst-case scenario from Peter's viewpoint. Such a situation occurs if all of Peter's supporters put Mary second on their list, but Mary's supporters place Peter last.

| a electors: | Peter > Mary > Paul |
| b electors: | Mary > Paul > Peter |

In this case, Peter receives $3a$ m-units from his supporters and b m-units from Mary's fans, who placed him last. Mary receives $3b$ m-units from her supporters, and another $2a$ m-units from Peter's voters who placed her second. In order for Peter to get elected by the Borda count, the number of m-units he is awarded ($3a + b$) must be greater than Mary's m-units ($3b + 2a$). Note now that $a + b = n$, the number of voters. Simple arithmetic then shows that Peter must garner at least two-thirds of the votes in a conventional majority election in order to guarantee his win according to the Borda count.

More generally, if there are n candidates, the winning candidate must receive at least $1 - 1/n$ parts of the votes cast in a simple majority election to be certain that he would have won even by the Borda count. (I derive this simple result in the mathematical appendix to this chapter.) In the case of two candidates that means receiving at least half the votes, which is the same as saying that a simple majority suffices. This makes sense. But a line-up of, say, five candidates would require a candidate to receive four-fifths, or 80 percent, of the votes to make him the undisputed winner of both election methods. This may seem overly stringent and it is. Most often less support suffices because the worst-case scenario usually does not arise.

An interesting case appears when there are more candidates than there are electors. In order for a candidate to obtain the threshold of $1 - 1/n$ parts of the votes, there must be at least n electors. If there are less than n electors, unanimity among the electors is required. (If there are six candidates but only five electors, the winner would have to receive at least five-sixth of the votes. This means he needs to obtain all five votes.)

There are problems with the Borda count, some minor, some major. One of the minor ones is that draws may occur. Borda did not express himself on what should be done if two candidates receive the same number of m-units. It may have been obvious to him that a runoff election would decide between the two. What if three or more candidates receive the same number of m-units? A second election by order of merit would be called for, and so on. And what about the case when an elector cannot rank two or more candidates because he is indifferent between them? Let us say there are five candidates and the elector ranked the first and second candidates, but is indifferent about the next three. Should they all receive three m-units, or one m-unit, or something in between?

A more serious problem is that, paradoxically, the winner of the Borda count may be nobody's favorite. It is easy to conjure up election results in which a candidate wins even though she is ranked no more than second best by all electors. For example:

11 electors: Paul > Mary > John > Peter
10 electors: Peter > Mary > John > Paul
 9 electors: John > Mary > Peter > Paul

Paul would get 63 m-units ($[11 \times 4] + [19 \times 1]$), Peter 69 ($[11 \times 1]$ + $[10 \times 4] + [9 \times 2]$), John 78 ($[21 \times 2] + [9 \times 4]$) and Mary, who is neither liked nor hated by anybody, would get 90 and win (30×3). The ranking would be Mary > John > Peter > Paul. By the way, a simple majority election would have given the ranking Paul (11 votes) > Peter (10) > John (9) > Mary (zero), the exact opposite of the Borda count.

Another paradoxical situation may arise through the sudden appearance of a clearly inferior candidate. Even though he would be ranked low on every voter's list, his addition to the roster may have a non-negligible influence on the election's outcome: the Borda counts of the front-ranked candidates could be changed, pushing a different winner forward. Let us assume that 51 electors prefer Ginger to Fred, and 49 prefer Fred to Ginger:

51 electors: Ginger > Fred
49 electors: Fred > Ginger

The Borda count declares Ginger the winner with 151 m-units ($[51 \times 2] + [49 \times 1]$), and Fred 149 ($[51 \times 1] + [49 \times 2]$). Now Bozo appears on the scene. Nobody really likes Bozo but his entry persuaded three of Fred's voters to rank Ginger even behind Bozo:

51 electors: Ginger > Fred > Bozo
46 electors: Fred > Ginger > Bozo
 3 electors: Fred > Bozo > Ginger

Now Ginger receives 248 m-units, Fred 249, and Bozo 102. Bozo's entry caused Fred to win.

Hence, by adding a dunce to the roster, the winner could be changed. The same may happen if a candidate drops out of the race or—may Heaven forbid—dies before the actual voting takes place. The most important challenge to the Borda count, however, is that it is open to manipulation through so-called strategic voting. This is the problem Pliny the Younger had been wrestling with (see chapter 2). We will have more to say about this practice in chapter 12.

Borda's suggestion was widely discussed in Paris. And the difficulties did not go unnoticed. Then another luminary appeared on the scene. His name was Marie-Jean-Antoine Nicolas de Caritat, Marquis de Condorcet.

BIOGRAPHICAL APPENDIX

Chevalier Jean-Charles de Borda

During several crossings of the Atlantic, Borda had the task of testing marine chronometers and studying methods to calculate the longitude of the ship's position. Toward the end of the eighteenth century, these were questions of paramount importance for maritime navigation. The latitude of a ship's position, that is, the distance north or south of the equator, can be ascertained relatively simply with the aid of a sextant or octant. Since these measurements are not affected by the earth's rotation, latitudinal positions can be determined by measuring angles of, say, the sun at noon over the horizon. Measurement of the longitude, though, is affected by the earth's rotation. Hence, a boat's east–west position cannot be established so easily. In order to ascertain the longitudinal position, a precise clock was needed that showed—wherever on the globe one may be—the local time at a reference point. Then, by comparing the reference time with the time at the current location, the ship's longitudinal position could be calculated. For example, if the sun at the current location is at its midday position and the clock, keeping the time of Le Havre, shows two o'clock in the afternoon, the navigator gathers that his ship is two hours, or 30 degrees, west of the port. (Twenty-four hours correspond to the full circle, that is, to 360 degrees. Hence, every hour's difference is equivalent to 15 degrees, which, along the equator is about 1,600 kilometers.) Together with the already determined latitude, the vessel's exact position on the globe is known. The clock's movement needed to be very precise, however. A deviation of just five minutes from the correct time at the reference point could translate into an error of up to 140 kilometers east or west. Many maritime disasters could have been avoided had exact timing devices been available to the ships' captains.

Pendulum clocks were of no use at sea, of course. They were meant to be hung on stable walls, not placed on vessels heaving and wallowing about in rough sea scapes. Watchmakers from various countries tried to invent timekeeping devices that would function sufficiently well even under extreme circumstances, but none were successful until the Swiss watchmaker Ferdinand Berthoud came to the rescue. He invented an isochronous balance wheel, driven by a spring that winds and unwinds at constant speed, which kept exact time even on boats rolling in foul weather. A first experiment showed that even after ten weeks of continuous operation the clock had accumulated an error of no more than one minute. In order to further test Berthoud's timepiece, King Louis XV ordered the mounting of an expedition. De Borda was appointed scientist in charge of the tests on board the *Flore*. The results exceeded the most optimistic expectations. After completion of the trip he and the ship's master composed a report titled "*Voyage fait par ordre du roi, en 1768 et 1769, dans différentes parties du monde, pour éprouver en mer les horloges de Monsieur Ferdinand Berthoud*" (Voyage undertaken by order of the king in 1768 and 1769 to different parts of the world in order to test the clocks of Mr. Ferdinand Berthoud at sea). The report was read to great acclaim at the Academy of Sciences. Berthoud was appointed the King's master watch-

maker and awarded a yearly pension of 10,000 francs.

During the American war of independence Borda was promoted to captain and put in charge of a vessel. Cruising in the Caribbean and along the American coast on board the *Seine* he participated in many exploits. In the famed Battle of the Saints, in 1782, six ships were under his command. It was to be the end of his career at sea, however. The British enemy proved stronger and after several hours of battle—his vessel disabled and a large part of his crew killed—Borda was taken prisoner. He was lucky, though. His captivity was not very severe and did not last very long. Upon his liberation he returned to France and was named director of the Engineering School of the French navy.

Only then did Borda, who was already fifty years old, start his second career as a scientist. It was to immortalize his name to a far greater extent than would his military exploits. At that time, great confusion reigned in all parts of France. Merchants, traders, and shopkeepers in every province and in every town used different weights and measures—which sometimes carried the same name. The confusion made commerce extremely difficult. In 1790 King Louis XVI set up a commission to study how the units could be standardized. Half a year earlier, the tentative proposal had been made to base measurements of lengths on the pendulum. The unit of measurement was to equal the length of the pendulum whose swing back and forth lasts exactly one second. The method seemed acceptable to Britain and the United

States, and French scientists were quite enthusiastic about the fact that their proposal was about to gain international approval. A proposal was submitted to the National Assembly, which referred it to the Committee on Agriculture and Commerce, which recommended it to King Louis XVI, who passed it on to the Académie des Sciences, which established a committee to further study the matter. Now things got into high gear. The blue-ribbon committee consisted of Paris's most celebrated scientists: Jean-Louis Lagrange, Pierre-Simon Laplace, Gaspard Monge, the foremost mathematicians of their time; Antoine Lavoisier, the great chemist; and the Marquis de Condorcet, mathematician, politician, and economist of whom we will read much more in the next chapter. Jean-Charles de Borda was named the commission's president.

The commission saw a few problems with the pendulum; for one, they felt that basing one unit of measurement (length) on another (time) was not an appropriate approach. After all, the division of the day into 86,400 seconds was artificial and could be changed at any time. In fact, Borda advocated dividing the day into 10 hours of 100 minutes each, the latter being made up of 100 seconds. The second problem was even more serious; since the earth is flattened at the poles the gravitational constant, which is responsible for the swing times, is not identical everywhere. Hence, at different places on earth, different lengths of pendulums are needed to produce a one-second swing. This problem could have been rectified by determining a specific place on earth where the pendulum would be timed and measured, but such a decision would have challenged the national pride of all countries who—it was hoped—would adopt the new system. Another reason to reject the pendulum was that time appears as a squared term in the equation that determines the period of its swing. The scientists wanted to keep everything simple and linear.

So another solution was sought. In the committee's first report, of October 1790, the members decided to adopt a decimal division of money, weights, and measures. Actually, the subdivision of units had not really been the issue, but the scientists found it nonetheless important to address the question, presumably because it was so convenient to use the ten fingers to count off the digits. This report was a prequel to their second report, of March 1791, in which the committee announced its decision to define the unit of length as the 10 millionths part of the quarter meridian, that is 0.0000001 of the distance from the North Pole to the equator. All that now remained was to measure this distance . . .

And this is where the real difficulties started. Measuring the earth in the late eighteenth century was a task comparable to building a space station in our days. The French scientists were not easily intimidated, however, and set themselves to the task. To facilitate the enormous undertaking, Borda invented a device that allowed the measurement of angles to a precision unheard of in his time. With this tool, measurements could be made by triangulating the landscape, and distances could be computed using trigonometry. But then the committee became aware of another difficulty; nobody had ever set foot on the North Pole, let alone measured any distance emanating from it. The scientists bypassed this problem by deciding to make do with the distance between Dunkirk and Barcelona. Measuring the distance between these two towns, establishing their latitudinal positions, and taking into consideration the earth's flattening at the North Pole, the total length of the quarter meridian could be computed.

But the revolution interfered. France was at war, the Republic was established, Louis XIV was tried and put to death, the Terror took over, Lavoisier was executed, Condorcet committed suicide or was murdered, the Academy was abolished. In short, confusion reigned. In the midst of all this, the scientists went about carrying out their task. One team of surveyors made its way south from Dunkirk—with their poles and flags and measuring devices—while another team worked northward from Barcelona over the Pyrenees. The members of the teams were arrested numerous times. More than once did they escape death only narrowly by pointing out that they were working on a replacement for the hated royal measuring system. Undeterred by all the hardships, they continued with their task until they met at the town of Rodez, about 500 kilometers south of Paris.

The undertaking had lasted nearly eight years. On November 28, 1798 the committee announced that the ten millionth part of the distance between the North Pole and the equator corresponded to 0.513243 *toises*, or, as we are wont to say nowadays, to one meter. Along with the liter and the gram, the meter became the official unit of measurement through a law enacted on December 10, 1799. Recent measurements, performed with the help of satellites, show that the French surveyors' measurement of the distance between Dunkirk and Barcelona was off by only about the length of two football fields. Thus, the meter that they had established more than two centuries ago was correct to within one-fifth of a millimeter.

MATHEMATICAL APPENDIX

Borda Count and Majority Elections

Let us assume that there are n candidates and E electors. a electors rank Peter first. If a is greater than 50 percent, Peter would be elected by the majority. Under which circumstances would he also be guaranteed victory by Borda's method?

In Peter's worst-case scenario, Mary would be ranked second by the a electors who ranked him first, and first by everybody else, that is, by $E - a$ electors. Peter would be ranked last by $E - a$ electors:

a electors:	Peter > Mary > . . . . . .
E – *a* electors:	Mary > . . . . . . > Peter

Peter would receive *n* m-units from each of the *a* electors who rank him first, and 1 m-unit from all others, for a total of *a* × *n* + (*E* – *a*) × *1*. Mary would receive *n* – *1* m-units from the *a* fans of Peter, and *n* m-units from all other electors, for a total of *a* × (*n* – *1*) + (*E* – *a*) × *n*.

For Peter's m-units to be greater than Mary's the following inequality must hold:

$$a \times n + (E - a) \times 1 > a \times (n - 1) + (E - a) \times n$$

Solving this, we obtain,

$$a/E > (n - 1)/n = 1 - 1/n$$

The term on the left-hand side, *a/E*, is the proportion of electors who place Peter first. If this proportion is greater than the right-hand side, *1* – *1/n*, Peter is guaranteed victory by the Borda count, even in the worst possible scenario.

CHAPTER SIX
THE MARQUIS

Scholarly debate in the French capital, with its newspapers, publishing houses, academies, and *salon* tradition, was always very lively. It was no different with Borda's voting scheme. As could be expected, his proposal of assigning points, or m-units, to preferences did not go unchallenged. The challenger came in the form of a nobleman, who was Borda's junior by ten years. His full name was Marie-Jean-Antoine Nicolas de Caritat, Marquis de Condorcet.

Born in 1743 in Ribemont, Condorcet was the only child of an ancient family of minor nobility. His father, a cavalry captain, was killed during a military exercise when Condorcet was only five weeks old. His mother, a fanatically religious woman, raised her son without any education. As a sign of devotion to the Virgin Mary and to the boy's childlike innocence, he was forced to wear white dresses until he was nine years old. This, his mother hoped, would guarantee her and her son's eternal salvation.

But then his uncle, a bishop, took over. Religion and devotion were all right, but even this man of the church thought this was going a bit too far. He hired a tutor so the boy could catch up with others his age, and then sent him to a Jesuit school in Reims, in the northern part of the country. Even though Jesuit schools were considered the best educational system Europe had to offer, they did not provide what one would nowadays consider a positive environment. Learning by rote and corporal punishment were the main instruments of instruction. Furthermore, rampant homosexuality among the monks and students left Condorcet with a hatred of the church that lasted throughout his lifetime. Nevertheless, he received a first-class education.

The boy's exceptional intellectual gifts soon became apparent and his uncle had him sent to the Collège de Navarre in Paris to continue his studies. In the first year, the college's program consisted of studies in philosophy, which Condorcet deeply disliked, and in the second of mathematics, at which he excelled. During his studies he had the good fortune of meeting the encyclopedist Jean le Rond d'Alembert. This celebrated mathematician and physicist had had a very unhappy childhood himself, born out of wedlock and abandoned by his mother upon his birth on the steps of a church. D'Alembert took the shy and awkward sixteen-year-old youth under his wings. Condorcet did not feel comfortable with the worldliness that reigned in the capital city. He was not good at speaking in company and would blush whenever spoken to. Nevertheless, he became a welcome guest at the salon of d'Alembert's companion, and possibly mistress, Julie de Lespinasse.

A first attempt to make a name for himself as a mathematician failed, because the results he had achieved were not new. But then, at age twenty-two, Condorcet published a work on integral calculus that was widely praised. Thus started his scientific career. Upon d'Alembert's recommendation he was elected to the Académie des Sciences four years later. Condorcet wrote more treatises, one of which was praised by his contemporary Joseph-Louis Lagrange, one of the leading mathematicians of the time, as a book "filled with sublime and fruitful ideas which could have furnished material for several volumes." After another four years he was elected the Académie's perpetual secretary. He had come to this position after taking to heart the advice given him by d'Alembert and a certain François-Marie Arouet, a.k.a. Voltaire. The two elder men had suggested to Condorcet that he gain experience in the most important skill the position required: writing obituaries for academy members who had passed away. In fact, the secretaries' *Éloges*, which covered all branches of the sciences, were not just simple summaries of scientists' lifetime achievements. They were more akin to learned chapters in the history of science than to the obituaries that we are accustomed to in today's newspapers. Actually Condorcet was no slouch in the literary realm, and in 1782 he was elected to the Académie Française, the highest literary honor to which a writer in France could aspire, again at the recommendation of his mentor d'Alembert.

While Condorcet was still preoccupied with his mathematical works he met Anne-Robert Jacques Turgot, a high official in the royal administration. Turgot, a brilliant economist, had a profound influence on Adam Smith, who lived in France at the time, and some of the ideas that Smith eventually incorporated into his *Wealth of Nations* came directly from Turgot. King Louis XVI named Turgot Minister of Finance in 1774. Looking for people that could be trusted, he, in turn, appointed his friend Condorcet Inspecteur Général des Monnaies, inspector general of the mint. (It is interesting to note that, across the Channel, the great Isaac Newton had held a similar position.)

Turgot, who saw the revolution approaching, realized the urgency of reform and the need to introduce competition and free markets into the French economy. Under the slogan "no bankruptcies, no new taxes, no debt" he attempted to make industry more efficient. He encouraged growth industries, abolished internal taxes on wheat, decreased government spending, curbed the extravagant outlays of the Royal Court, and ended the guild system, which had held a stranglehold over commerce and industry ever since the Middle Ages. All this did not sit well with the established orders, and after a while Turgot had made enemies of just about everyone in France. Nevertheless, for as long as the king supported him, he was safe. But then Turgot committed a grave error. Citing the need for economic belt-tightening he refused some favors to Queen Marie Antoinette's protégés. With that he broke one of the most important laws at Court: don't mess with the king's wife. Turgot was dismissed.

The Swiss banker Jacques Necker succeeded him at the ministry. He proceeded to reverse most of his predecessor's policies, which was one of the reasons for the eventual outbreak of the revolution. With his protector gone, Condorcet tendered his resignation. But the king refused and Condorcet stayed on at the mint for another fifteen years. Condorcet continued to write learned tracts on mathematics, economics, political science, and human rights.

At age forty-three Condorcet fell madly in love with Sophie de Grouchy, a lady more than twenty years his junior. The eldest daughter of the Marquis de Grouchy, a former page of Louis XV, she was said to be the most beautiful woman of her time in Paris. Condorcet and the young lady were of one mind in all their ideas and made an ideal couple. They were married in 1786. As was the custom for intellectual women in Paris, Sophie kept a salon at the couple's residence, the Hôtel des Monnaies. One of the guests who frequented these gatherings was an American by the name of Thomas Jefferson. Apart from organizing her salon, Sophie kept herself busy translating the works of Adam Smith—another of the guests at her salon—into French. She was also an accomplished portraitist, a skill that would become her sole means of support when hard times befell her. Four years after they got married the couple had a daughter, whom they named Eliza. Condorcet was a loving husband and a doting father.

As both president of the Académie des Sciences and member of the Académie Francaise Condorcet was by now one of France's foremost intellectuals. A true man of the enlightenment, he championed every liberal cause he could think of: economic freedom, tolerance toward Protestants and Jews, legal reform, public education, abolition of slavery, equality of all races. For example, he argued eloquently for women's rights. "Why should beings exposed to pregnancies and to passing indispositions not be able to exercise rights that no one ever imagined taking away from people who have gout every winter or who easily catch colds? . . . It is said that women have never been guided by what is called reason despite much intelligence, wisdom, and a faculty for reasoning developed to the same degree as in subtle dialecticians. This observation is false."

When turmoil broke out in 1789 Condorcet could not sit back. Leaving mathematics behind, he took a leading role in the revolution and was elected to the legislative assembly as a representative of Paris in 1791. Belonging neither to the more radical Montagnards nor to the more moderate Girondins, he tried to mediate among the various factions and temper the more extreme elements. As one of the more reasonable men in the legislative assembly, Condorcet was chosen to draft a constitution for the new nation. When the assembly was replaced by the convention a year later, Condorcet felt closer to the Girondins, who had in the meantime lost their power to the Montagnards. The latter, led by Robespierre, put down any opposing opinion with an iron fist, abolished royalty and put King Louis XVI on trial. Condorcet supported the trial but opposed the death penalty, a stand that did not make him a favorite of the Montagnards. It did not help the king either, who was executed on January 21, 1793.

Condorcet was not a speaker who could sweep away his audience. His rhetorical skills had not improved much since his youth; he was still shy and his voice did not carry. It was, therefore, not surprising that when he introduced his draft constitution to the assembly, he was unsuccessful. When his opponents presented their version of a constitution, an adulteration of his own initial draft, Condorcet protested against it with all his might and was promptly accused of being a traitor. I recount the tragic events that followed in the additional reading section.

The Marquis de Condorcet, one of the most remarkable men the French Revolution produced—politician, constitutional lawyer, mathematician, writer—left a great legacy. Important works in mathematics intermingle with texts on social issues. Some of his most intriguing texts, both as politician and as mathematician, were the contributions to the theory of voting and elections. His name is associated until today with one of the great social puzzles of all times: the Condorcet Paradox.

The paradox, to which we alluded in the previous chapter when talking about Jean-Charles de Borda, refers to a serious shortcoming of majority decisions. It is universally believed that decisions should be made, differences of opinion decided, judgments rendered, and officials elected, by taking votes and then counting which of the alternatives garnered most support. In fact, majority decisions represent one of the pinnacles of democracy. After all, the tenet of "one man one vote" rests on the assumption that the majority is always right. But to the surprise of many of Condorcet's, and our, contemporaries this assumption is fundamentally flawed. The Marquis showed that majority opinions are not always what they purport to be.

In 1785 Condorcet wrote a two-hundred-page pamphlet titled *"Essai sur l'application de l'analyse à la probabilité des décisions rendues à la pluralité des voix"* (Essay on the application of probability analysis to majority decisions). He dedicated the work to Turgot, saying that it was he who had taught him that political science is amenable to the same degree of certainty as mathematics. As a case in point, Condorcet chose to demonstrate the power of mathematics by applying it to decisions made by majority votes. His analysis is not applicable only to citizens electing their leaders, he wrote, but also to judges in courts of law who must decide between guilt and innocence of an accused.

The essay was published fifteen years after Jean-Charles de Borda's address to the Académie des Sciences about ballot elections, and four years after the publication of his proposal to assign points, or m-$units$, to a candidate according to his ranking. Condorcet acknowledged Borda's contribution in a footnote saying that it had came to his attention only when his essay was already being printed.

As we shall see in the next chapter, Condorcet and Borda did not get along very well, but they did agree on one issue. Both had a somber view of majority decisions. Unlike Llull and Cusanus, who firmly believed that majorities reveal God's will and absolute truth, the two Frenchmen did not think that majority decisions were ipso facto correct decisions. Condorcet was convinced that societies had adopted majority rule for a much more pragmatic reason. Subordinating individuals to the will of the majority was meant to safeguard peace and quiet. Authority had to be placed where force is, and force is on the side on which most of the votes come down. Hence, for the good of the people, the will of the smaller number had to be sacrificed for the will of the larger in order to keep everybody quiet.

To buttress his claim, Condorcet cited instances from ancient times. The Romans and Greeks did not necessarily seek truth and try to avoid errors. What they strove to do was to balance the interests and passions of the various factions that made up their states. Whenever decisions were made, whether just or unjust, true or erroneous, reasonable or unreasonable, they had to be sustained by force. And since force is wielded by the majority, even incorrect decisions were adopted, if only they enjoyed the support of the majority. Subjecting decisions to tests of justice, truth, or reason would have put unnecessary restraints on the faction's authority. So might was right after all.

But eventually methods were sought that would permit decisions based on reason. The search for mechanisms less prone to error had started a long time before the age of enlightenment. During the centuries of deepest ignorance, a certain unease with majority decisions had already surfaced, especially when dispensing justice. Probably the most important problem a court of law faces is that judicial errors can result in people being convicted of crimes they did not commit. Therefore, attempts had been made in the Middle Ages to give the courts a form that would increase the probability that their decisions reflect the truth. Distrust for rulings in which one judge tipped the scales led the French to demand more than simple majorities to convict an accused. In England unanimity was required whenever juries rendered a decision. The Catholic Church's court of appeals demanded no less than three unanimous rulings in order for a judgment to be valid. (Of course, witch trials required neither a simple nor a qualified majority. Truth was elicited by the proven method of subjecting the poor women to various forms of torture.)

Condorcet gave a taste of how mathematical ideas, more specifically probability theory, could be applied to decisions rendered by courts of law. He pointed out that the requirement of a plurality more stringent than a majority of one would render miscarriages of justice less probable. The larger the plurality required in a court, the smaller the probability that an innocent man would be convicted. This immediately begged the question, how far this certainty should be carried. And it raised a second problem: guilty defendants should not be declared innocent simply because the majority requirements had been carried too far.

So, after his rather pessimistic, if pragmatic, view of the advantages of majority decisions, Condorcet delved into the disadvantages. At the outset of his essay he apologized to mathematicians who would find the mathematical methods only of limited interest. Indeed, nothing more than basic arithmetic is needed to follow Condorcet's reasoning.

It is on page sixty-one of his essay that Condorcet presents the reader with the famous paradox. He illustrates it with an example of sixty voters who have to elect one of four candidates to a certain position. Here I give a simpler example with just three voters. Let us say Peter, Paul, and Mary must decide what to buy for their after-dinner drinks. Peter prefers Amaretto to Grappa, and Grappa to Limoncello. Paul prefers Grappa to Limoncello, and Limoncello to Amaretto. Finally, Mary prefers Limoncello to Amaretto, and Amaretto to Grappa.

Peter:	Amaretto > Grappa > Limoncello
Paul:	Grappa > Limoncello > Amaretto
Mary:	Limoncello > Amaretto > Grappa

Committed as they are to democratic values, the three diners decide to go by the majority opinion. They take votes and the preferences become quickly apparent. A majority prefers Amaretto to Grappa (Peter and Mary) and a majority prefers Grappa to Limoncello (Peter and Paul). Based on these two rounds they can make their decision: purchase a crate of Amaretto.

But surprise, surprise: Paul and Mary protest. What happened? The most reasonable selection method was used—one person one vote —and they still aren't happy? Do they want to change the rules in mid-game? Well, they have a legitimate grumble. Paul and Mary point out that they would prefer even Limoncello, the lowest ranked option, over Amaretto. How come? Here is the clincher: had the three campers had a third round of voting, between Limoncello and Amaretto, a majority would have preferred Limoncello (Paul and Mary). So let them buy Limoncello and get it over with. But wait a minute. Buy Limoncello, and Peter and Paul—yes, Paul, the guy who insisted on the third vote because of his dislike of Amaretto—will protest just as vigorously. They prefer Grappa to Limoncello. So here we have it, a paradox. One does not argue about tastes and Peter, Paul, and Mary have perfectly reasonable preferences. Try as you might, the final result is that Amaretto is preferred to Grappa, Grappa to Limoncello, Limoncello to Amaretto, Amaretto to Grappa, Grappa to Limoncello. . . . We could go on and on.

So what is the solution? The depressing answer is that there is none. There is no way out of Condorcet's Paradox. Whatever the choice, a majority always prefers a different option. Preferences cycle through all the alternatives and the paradox persists. The fact that the majority prefers Amaretto to Grappa, and Grappa to Limoncello, simply does not imply that Amaretto is preferred to Limoncello by the majority. In mathematical lingo, one says that "majority opinions are not transitive." What a letdown for democracy.

Condorcet's Paradox can be the source of much abuse. For example, a person setting the agenda at a board meeting can subtly influence the outcome of decisions by manipulating the order in which votes are taken. Let us say a company wants to reward its employees by installing a cafeteria, a health club, or a nursery on its premises. A decision must be made, and the company's CEO charges the personnel director with organizing a board meeting. The personnel director hates sweaty basements and has no predilection for screaming kids. She does, however, enjoy taking time off from her busy schedule for the occasional cup of coffee. At the meeting she takes the first two votes, and the cafeteria comes out on top. Since small talk and prevote discussions, stoked on by the personnel director herself, have taken up most of the morning, there is little time left for any more votes. Some board members need to use the bathroom facilities, others want to have a smoke, and anyway lunch is waiting in the executive dining room. "Let's just break up now," the sly operator may say. "Since the cafeteria is preferred to the health club and the health club to the nursery it is obvious that the majority wants a cafeteria." Nobody bothers to find out whether the nursery would have bested the cafeteria if a direct vote had been taken. And this is how the personnel director can have her way.

Hence with good reason the deeply troubled Condorcet feared that the paradox poses great dangers. Since ignorant masses could be manipulated by corrupt politicians and charlatans he decided that the people had to be informed about their rights and obligations as citizens. If a society's philosophers did not find the courage to enlighten the unsuspecting people, tyranny could set foot in the country and maintain itself. Condorcet, who devoted himself to seeking truth and to serving the fatherland, took that task upon himself.

As a vehicle for his educational efforts, Citizen Condorcet (it was no longer fashionable—in fact it was downright dangerous—to carry the title Marquis), together with Citizen Sieyes and Citizen Duhamel founded the *Journal d'Instruction Sociale* (Journal of Social Education), a weekly publication that would devote its pages to educating the public about their rights and duties. One should never forget, the editors reminded the readers in the prospectus for the new journal, that while liberty and equality were the most important assets of an enlightened people, they could also be the cause of the greatest harm if, due to ignorance, the people did not know how to safeguard these assets. The journal's aim was not to lecture its readers, the editors stressed. The objective was to enable them to form their own opinions.

The journal was to be launched in 1793 and be published every Saturday. Whatever profits there would be, the editors promised, would go to the National Institute of the Deaf and Dumb—the hearing impaired in modern parlance—at whose premises the journal was to be printed. The journal's first issue appeared on June 1, 1793, all of its three articles having been written by Condorcet. After a philosophical investigation into the meaning of the newly coined term *revolutionary* and an essay on progressive taxation, the booklet closes with the eight-page paper that is of primary interest to us. It is titled "*Sur les élections*" (On elections). In it Condorcet outlined his ideas on the electoral process.

The French were about to grant themselves a constitution that would decide the nation's fate. Would the people be governed by reason or by intrigue, by the will of all or by the will of a few? Would liberty be peaceful or agitated? The answer to these questions, the very survival of a well-functioning society, depended on the quality of the popular choices, Condorcet wrote. Constitutional shortcomings themselves pose no immediate dangers. As long as honest, publicly minded men rule the country—in spite of his avowed feminism Condorcet did not go so far as to include women—there would always be occasion to correct any threats to the nation that may arise. If corrupt men take over, however, even the best laws become only feeble ramparts against ambition and intrigue.

But honest people who base their elections, judgments, and decisions on a plurality of votes can be led into the feared cycles. So if majority decisions are not the solution, what is? Condorcet thought long and hard about the problem and finally came up with a suggestion. As in his essay of 1785, he recommends combinatorial methods and probability theory as the surest way to avoid the ills of conventional methods. His suggestion will sound strangely familiar to the readers of this book. It is very similar to the method that Ramon Llull had already proposed 500 years earlier.

When an elector chooses a candidate for a certain post, he performs a series of judgments. He does so by comparing all possible pairs of candidates, examining the reasons to vote for the one or the other, weighing them, and then expressing a choice. By doing this for all pairs of candidates, he obtains a ranking and the top-ranked candidate is the individual's favorite for the job. If not all electors express complete rankings—either because they are indifferent between some candidates or because some of them are unknown to them—the election outcome may not reflect the true preferences of the assembly. Nevertheless, Condorcet cautions, electors should not be forced to choose among candidates they do not know because this would simply result in a random ranking. Rather, he recommends—like Llull did half a millennium previously—that a list of acceptable candidates, well known to all electors, be drawn up before the voting starts. The electors must then make a complete listing (including indifferences) only among those deemed eligible.

If the elector suddenly realizes that he prefers Alexander to Bertram, Bertram to Charles, and at the same time Charles to Alexander, then, Condorcet asserts, at least one of the choices must have been based on an erroneous assessment of the relative strengths of the candidates. (Condorcet tacitly assumes that choices must accord with common sense and therefore be transitive.) In this case the elector will reexamine his set of choices and clean it of the ones that led to inconsistencies. Reevaluating all judgments, he identifies those that may lead to the absurd situation, and drops, or reverses, the one that he deems most improbable. For example, if he strongly prefers Alexander to Bertram and strongly prefers Bertram to Charles, while his preference of Charles over Alexander is only slight, he will drop the latter preference. Thus the elector will end up with a complete ranking of the candidates.

Now comes the actual election. An election by an electoral college is the aggregation of the individual rankings. Condorcet suggests that after the electors have made up their minds about the candidates' relative positions, they get together to judge the candidates. Each candidate is paired against each of the other candidates in a series of showdowns. The electors express their preferences, and the candidate who receives more votes is considered to be superior to the other. After all pairings have been performed, the candidates are ranked. In the final list, a candidate who won the showdown against a particular competitor will be ranked above him and the person who comes out on top of the list will be declared the winner.

In the ideal situation, where the best candidate wins all contests against the other candidates, an unambiguous winner exists; he will be declared the "Condorcet winner." (A candidate who loses all showdowns against the other candidates is called a Condorcet loser.) But things are not usually as simple since, in general, no unambiguous list can be drawn up. A Condorcet winner, the ideal candidate, superior to every other contestant, exists only rarely. Usually no candidate wins every single showdown. Even an exceptionally strong contestant is bound to lose some of them. This, of course, produces cycles, and when that happens, no Condorcet winner exists. What is to be done?

When cycles appeared in an individual's preferences, Condorcet argued that the situation does not accord with common sense and one of the preferences should be reversed. But there is an important difference between a single man's preferences and an election. As was shown in the Grappa-Amaretto-Limoncello example, when aggregating the preferences of three or more electors, cycles may occur even if the individual electors' judgments are perfectly consistent. So Condorcet suggested that an assembly use the same method to resolve a cycle as an individual does when examining his choices: at least one of the showdown results must be deleted. But which one? It cannot be argued that preferences expressed by the electors were unreasonable. After all, they were based on a majority of votes. Condorcet saw a way out: the preference with the feeblest majority is to be dropped.

So Condorcet's proposal entails the two-by-two showdowns with which we are already familiar due to Ramon Llull's work in the thirteenth century. But there is an important difference. Llull had advocated the election without further ado of the candidate who won most of the contests. Condorcet, in contrast, proposed to check the whole ranking, all the way to the bottom, for inconsistencies. If the top-ranked candidate turns out to be inferior to another candidate, the result of the showdowns that produce the cycle should be dropped. In the end, the Condorcet winner may not be identical to the winner according to Llull.

Let us analyze an election for a monastery's prior. Eleven candidates vie for the post. The best result was achieved by Brother Angelo who won the showdowns against all competitors, except for the one against Brother Giulio. He thus gets nine points. Brother Giulio who won all showdowns except the ones against Brother Innocenzo and one other competitor came next with eight points. Brother Innocenzo, who lost against Brother Angelo, won against Brother Giulio, and won another six showdowns, came in third with seven points. The following table summarizes the results.

TABLE 6.1

	Showdown against:				
	Angelo	Giulio	Innocenzo	. . . .	Total
Angelo	-	loses	wins	8 additional wins	9 points
Giulio	wins	-	loses	7 additional wins	8 points
Innocenzo	loses	wins	-	6 additional wins	7 points

Based on the total score, Ramon Llull would have had Angelo elected prior. But Angelo was beaten by Giulio. And Giulio was beaten by Innocenzo, who in turn was beaten by Angelo. We have a cycle. Who should become prior?

Condorcet suggests scrutinizing the election results more closely. Let us say that Angelo lost the crucial showdown against Giulio with a whopping 1 vote to 8, while Giulio lost his showdown against Innocenzo very narrowly with 4 votes to 5. Finally, Innocenzo lost his duel against Angelo with 2 votes to 7:

TABLE 6.2

	Showdown against:		
	Angelo	*Giulio*	*Innocenzo*
Angelo	-	loses 1:8	wins
Giulio	wins	-	loses 4:5
Innocenzo	loses 2:7	wins	-

According to Condorcet's proposal, Brother Giulio's loss, being the narrowest, would be deleted from the tally. Hence, the cycle would be broken and Brother Giulio would become the new prior.

What happens when several cycles appear, as often happens when more than three candidates present themselves? In this case inconsistencies are even more probable. But Condorcet sees no special problem. If dropping one preference does not fix the cycles, the majority must have erred more than once. The solution consists in dropping as many judgments as needed, until an unambiguous winner can be determined. All judgments that lead to inconsistencies are reexamined and dropped one by one, starting with the ones that have the narrowest majorities.

The cycle-breaking mechanism Condorcet proposed seems like a good idea. So why not use it? The problem is that it is not easy to implement. With, say, ten candidates there are forty-five judgments. (The first candidate meets nine competitors, the next eight, and so forth. More generally: with n candidates there will be $n(n-1)/2$ showdowns.) It is no simple task to single out those judgments—among the forty-five—that lead to the inconsistencies. But an even more serious problem can arise if two or more inconsistent judgments obtain equal majorities. Let us look again at the simplest example, the after-dinner drinks. Each choice has the same two-to-one majority. So where should the cycle be broken? Which showdown should be dropped? The one between Amaretto and Grappa, the one between Grappa and Limoncello, or the one between Limoncello and Amaretto?

Condorcet's proposal seems a very reasonable method, but in its purest form it is fairly useless. Of course a Condorcet winner, the candidate who beats all others, would be the preferred winner. If there are only two contestants, the outcome is obvious: the one who beats the other is the Condorcet winner. But even one additional candidate may lead to a messy situation, as witnessed by the Amaretto-Grappa-Limoncello example. The more candidates there are, the more unlikely it is that a Condorcet winner exists. And then, there is the large number of showdown contests that would have to be performed, a near impossible task even for a moderate number of candidates.

*   *   *

After centuries of making do with majority elections, all of a sudden the method was shown to be defective. And now there were not one, but two new proposals, both of which had their advantages and disadvantages. In Condorcet's two-by-two showdowns an inferior candidate would never be elected, but there was no guarantee of a winner. Borda's m-units-for-rank scheme takes the electors' true preferences into account, but the eventual winner could very possibly turn out to have been nobody's favorite. And if there are Borda and Condorcet winners, the two may not be identical. Paradoxes abounded everywhere. Neither of the methods was undisputedly superior to the other. This did not keep the two savants from flaunting the advantages of their own election method while putting down the other.

Nevertheless, both men deserved great honor and great honor they received: in Paris's 9th arrondissement a street was named after Condorcet, and in the 3rd we find a street named after Borda. The honors did not end with the rue Condorcet and rue Borda either. To underscore their international, and even outer-spatial reputations two lunar craters on the moon have been named after Borda and Condorcet. For good measure there is also a Cusanus Crater, but nothing on the moon has so far been named for Ramon Llull.

BIOGRAPHICAL APPENDIX

Marquis de Condorcet

Branded a traitor and fearing for his life, the Marquis de Condorcet took refuge in the house of a devoted woman, Madame Rose Vernet. This lady, a widow who supported herself by renting out rooms in her house in the rue des Fossoyeurs, was a person of exceptional character. Her name would be unknown to us today, had it not been for the exceptional courage she displayed during the great Terror by sheltering the wanted fugitive.

Only two of Mme Vernet's tenants knew of Condorcet's identity. One of them was a Montagnard by the name of Marcoz, who was told of the secret but kept it and, in fact, provided Condorcet with newspapers and information about the developments in the outside world. The other tenant who was in on the secret was Mme Vernet's cousin, the mathematician Sarret. The only other person who knew of Condorcet's identity was Mme Vernet's loyal servant, Mademoiselle Manon. Shared secrets and the cramped quarters made for a romantic atmosphere, and according to some accounts Mme Vernet and Monsieur Sarret eventually got married.

While Madame and Monsieur may have had a budding romance, Condorcet was very lonely. The only contacts he had in his hideout were with Sarret, Marcoz, and Mme Vernet. Occasionally Sophie came to see her beloved husband, but visits were dangerous and therefore rare. He never got to see his four-year-old daughter again, whom he so dearly loved. A letter that he sent Eliza touches the readers' hearts even two hundred years later: "Whatever the circumstances in which you read these lines, which I am writing far away from you, indifferent as to my own fate but preoccupied by yours and your mother's, remember that nothing can guarantee that those circumstances will last. Get into the habit of working, so that you are self-sufficient and need no external help. Work will provide for your needs; and though you may become poor, you will never become dependent on others. ... My child, one of the best ways to ensure your happiness is to preserve your self-respect, so that you can look back on your whole life without shame or remorse, without seeing a dishonorable act, nor a time when you have wronged someone without having made amends. ... If you want society to give you more pleasure and comfort than sorrow or bitterness, be indulgent and guard yourself against egoism as a poison which ruins all its pleasures. ... "

In order not to arouse suspicion concerning the whereabouts of her husband, Sophie made one of the hardest decisions of her life; with his consent she divorced the Marquis. By then Sophie was nearly penniless and had to support herself and her daughter by drawing portraits. Actually portrait drawing was quite a good business in these uncertain times. Many Parisians, not knowing what the future may bring, wanted to leave their likeness to their next of kin.

Condorcet spent a cold and lonely winter working in solitude on his last text, *Esquisse d'un tableau historique des progrès de l'esprit humain* (Sketch for a historical picture of the progress of the human mind). The partly imaginary description of the human race's progress from savagery to a future state in which equality among classes and nations would reign, and human nature would be perfected, was to become Condorcet's legacy.

After staying at Mme Vernet's boarding house for five months, the Marquis had reason to fear that his hideout in the rue des Fossoyeurs (renamed today rue Servandoni) was being staked out. An unknown man had shown up at Mme Vernet's doorstep, on the pretext of wanting to rent a room. He had asked strange questions and then left again. Condorcet felt that it was no longer safe to stay in this hideout. Had he been discovered, it would have meant the guillotine not only for him but also for Mme Vernet, and his wife's life would not have been spared either. Against the express wishes of his devoted landlady, Condorcet left the boarding house.

Dressed as a commoner, he set out for the house of Amélie and Jean-Baptiste Suard, close friends of his from better times. In the countryside outside Paris, he hoped to receive temporary shelter. It was a perilous journey. Condorcet managed to cross the city lines where only six days earlier a former member of the convention by the name of Masuyer had been recognized, tried, and immediately executed.

After a long and strenuous walk—Condorcet had not been able to exercise his legs for nearly half a year—he finally arrived at the Suards' home. A maid opened the door only to tell him that her master and mistress had left for Paris that same morning. The lonely fugitive spent the next two days without food, wandering around and sleeping under the open sky. When his friends finally returned from Paris they were too afraid to take him in. One could hardly have blamed them. Giving refuge to a wanted man was punishable by death, and the less than trustworthy maid had taken a close look at the unshaven stranger. Suard promised that he would try to obtain a passport for him, and Condorcet left the house.

He took refuge at a country inn trying to blend in with the locals. But his noble demeanor soon betrayed him and when he ordered an omelet with an "aristocratic amount of eggs"—believed to have been 12—

his cover was definitely blown. Asked to identify himself, Condorcet, who carried no papers, tried to pass himself off as a chamber valet by the name of Pierre Simon. Pending verification of his identity, the unknown man was put in a prison cell. Two days later he was found dead. The cause of his demise was never determined. Did he die of natural causes, did he commit suicide, or was he murdered because he was too popular in Paris to be executed? One version has it that a friend of his, a medical doctor, had given him a vial of poison a long time ago that Condorcet kept hidden in a ring on his finger. It was to spare him the guillotine if and when things should go terribly wrong. Had he used it? We will never know.

Sophie was also arrested but soon set free. She survived her husband by twenty-eight years. When Eliza was seventeen, she married an Irish general, twenty-seven years her senior, by the name of Arthur O'Connor. O'Connor had been an indefatigable fighter for Irish independence. Arrested by the English and kept in prison for five years, he finally agreed to exile in France and became a Général de Division under Napoleon. He and Eliza bought an estate south of Paris where they raised three sons who tragically died. (Other sources indicate that Eliza and O'Connor's offspring served as officers in the French army.) After his retirement from the army O'Connor became a prolific writer on social and political subjects, even helping to edit the twelve volumes of Condorcet's works.

THE MATHEMATICIAN

When writing his essay in 1785, Condorcet was apparently already aware of Borda's contribution of 1781. He admitted as much in a caustic footnote in which he acknowledged that the existence of Borda's paper had been pointed out to him by friends, at a time when his own paper was already being printed. Somewhat patronizingly, he claimed that he would not have known anything about the paper save for the fact that some people had mentioned it to him. As is now believed, however, Condorcet was being less than truthful. In 1781, he was the perpetual secretary of the Académie des Sciences, and as such was responsible for editing the academy's *Mémoires*. He could not have been unaware of what was being printed. More likely, he himself decided to publish Borda's paper.

Condorcet did not think highly of Borda. In fact, he did not even consider him a very capable mathematician. While undeniably talented, Condorcet said, Borda had to take recourse to the lesser science of engineering, building ships and fortifications, after failing at mathematics. According to Condorcet, Borda was not even qualified to be a member of the academy of sciences and had gained entry to this hallowed temple of scholarship not by erudition or scholarship but by wish of the king. In a letter to a friend, Condorcet wrote that Borda likes to talk a lot and wastes his time tinkering with childish experiments.

Why would Condorcet, the self-styled gatekeeper of French science, publish a supposedly inferior paper? Actually he did even more than have it published. Borda's paper was prefaced by a highly complimentary review, and according to custom it was the editor who wrote the prefaces. So Condorcet, who considered Borda unworthy of even being a member of the academy, not only published his paper but also sang its praises. Why did he do that? Had he not recognized Borda's method as a challenge to his own? Whatever the motives, publishing the paper gave him occasion to put forth his own—allegedly superior—election method.

Condorcet attacks the Borda count by means of an example. Without even mentioning Borda, except for a caustic reference to "a famous mathematician," he attacks his method with the following situation. Eighty-one voters are to make their choices between Tom, Dick, and Harry. (The numbers are Condorcet's, the candidates' names are mine.) Their preferences are as follows:

30 voters:	Tom > Dick > Harry
1 voter:	Tom > Harry > Dick
10 voters:	Harry > Tom > Dick
29 voters:	Dick > Tom > Harry
10 voters:	Dick > Harry > Tom
1 voter:	Harry > Dick > Tom

A short explanation follows, in which Condorcet explains that the Borda count awards three units of merit for the top, two for the middle, and one for the bottom rank. Hence, Tom receives 182 m-units, Dick 190, and Harry 114. Accordingly Dick is declared the winner. But does this choice reflect the true will of the voters? Delving deeper into the preferences—after all, this is what the Borda count is about—one realizes that 41 voters prefer Tom to Dick, while only 40 voters rank Dick above Tom. Gleefully, Condorcet points out that Borda would have Dick win, even though less than half of the voters prefer him. So who should be elected?

Condorcet comes down clearly on Tom's side, who, after all, is the Condorcet winner. The example, which to Condorcet's credit is not at all contrived, proves that the Borda count may fail to elect a candidate like Tom who would beat all other competitors in individual showdowns.

As was already pointed out, there is another problem with Borda's m-unit method. Let me bring to mind the situation. There are five electors who must choose between Laurel and Hardy. Three of them prefer Laurel, two prefer Hardy:

| 3 voters: | Laurel > Hardy |
| 2 voters: | Hardy > Laurel |

The three voters would give Laurel two points each, and two voters would give him one each, for a total of eight. Similarly Hardy would receive seven points and lose. Now Goofy enters the game and the electors rank the candidates in the following manner:

| 3 voters: | Laurel > Hardy > Goofy |
| 2 voters: | Hardy > Goofy > Laurel |

This time Laurel receives eleven m-units, Hardy twelve, and Goofy seven. Hardy, the former loser, wins. Even though Goofy had no chance of winning, his participation in the race reversed the outcome. If he does not participate, Laurel wins, if he does, Hardy wins.

When Condorcet's essay appeared, Borda was busy attending to military matters, and it is not known whether he ever responded to the criticism. Actually there was no need to do so, because he had anticipated it in his paper. He had mentioned two-by-two contests in his original paper but dismissed the method by saying that it would require too many showdowns. Little did he know that Condorcet would present exactly this method as his preferred electoral scheme.

At about that time, the Academy was deliberating a revision of its bylaws for the election of members. Whenever a *pensionnaire* died, a new member had to be elected to the vacant seat from among several candidates. Adoption of Borda's method was being considered. But the shortcomings were obvious. The introduction of a new candidate, even if he was not a serious contender, would change the Borda count of all candidates and—worse still—could do so in differing manners. The discourse about the relative merits of the two election methods was raging in full swing when a third heavyweight entered the ring: Pierre-Simon de Laplace.

This mathematician, born in 1749 in Beaumont-en-Auge, Normandy, was one of the members on Borda's committee that was charged with standardizing weights and measures. His father was a businessman. His mother came from a family of farmers and landowners. Pierre-Simon attended a Benedictine school and then went on to study at the University of Caen. He started out studying theology and initially wanted to embark on a career in the church. But inspired by two teachers, he soon discovered his love for mathematics. Unable to further his education beyond their own limited abilities, they encouraged him to go to Paris, by now the usual destination in France for talented young men. They gave him a letter of introduction to Jean d'Alembert. The esteemed scientist was sufficiently impressed by the visitor to recommend him to the École Militaire, where Laplace would teach geometry, trigonometry, and elementary calculus to the adolescents of the Parisian middle class.

He read thirteen papers before the academy, ranging from differential equations and integral calculus, to probability theory, celestial mechanics, and the theory of heat. The subjects he talked about before the learned audience were to occupy him all his life. But even though d'Alembert and Condorcet remarked on the quality of his work, he was turned down twice for membership to the Academy of Sciences. Finally, in 1773 he was elected an adjunct member of the prestigious institution.

In 1784 Laplace was appointed examiner in the Artillery Corps. In this capacity he examined, and passed, a sixteen-year-old cadet from Corsica by the name of Napoléon Bonaparte. How history might have changed had he failed him! Laplace soon considered himself the best mathematician in France and let that fact be known to everyone. Although he was, arguably, correct in this assessment, it did not endear him to his peers. His relationship with d'Alembert also deteriorated because he thought his former benefactor's findings were obsolete.

In August 1793 the revolutionaries closed the doors of the Académie des Sciences since the furthering of human knowledge was no longer to be a privilege of the elites. Priority would be given to the instruction of the general public. But in 1796 the Academy resumed its activities and one question of paramount importance was how new members should be elected. The Borda rule was chosen, in spite of its shortcomings. For a few years, everybody was reasonably happy, until one of the newer members complained. And when this particular academy member had anything to say, the others had better listen; it was Laplace's former pupil, the cadet, and by now general, Napoleon Bonaparte. Actually, the academicians did not have to listen to Napoleon's utterances too often. It is said that this was the only time he ever rose to speak as a member of the academy.

Napoleon had a valid point. Criticism had already been leveled against Borda's method five years earlier by Pierre-Simon Laplace. What was it that Laplace and Napoleon found unsavory about the Borda count? They realized that the method of "voting by merit" was open to a subtle sort of manipulation that has since come to be called "strategic voting." Laplace first voiced his criticism in his lectures at the newly founded teachers' college École Normale Supérieure (ENS) in the spring of 1795. The lectures appeared in written form in 1812, in the same year that his much bulkier and far more detailed treatise *Théorie analytique des probabilités* was published, in which he repeated his analysis.

Altogether Laplace gave ten lectures at ENS. The course started easily enough: how numbers are formed and written, how they are added, subtracted, multiplied, and divided. In fact, the first lecture was downright stultifying, except for the remark that the German mathematician Leibniz had introduced binary digits—nowadays known as bits in the context of computers—as a numbering scheme in which the one denoted God and zero stood for "nothing." In succeeding lectures the professor quickly got down to business. In rapid succession he dealt with equations of higher order, imaginary numbers, transcendental equations, algebraic geometry, and many more subjects. No wonder the poor students who wanted simply to get a grounding in mathematics so they could make a living as schoolteachers found the going rough. The tenth, and last, talk was devoted to probability theory.

Laplace opened this lecture with a wish list of subjects that he would have liked to discuss had there been more time: differential equations, integral calculus, mechanics, and astronomy. But given the short time at his disposal, all he could do was refer the students to his forthcoming book *Exposition du système du Monde*. One suspects that the auditors were not all too unhappy to be spared additional lectures on difficult subjects, important as they may be.

He then introduced the subject—on which he would elaborate in a major treatise seventeen years later—with the remark that the theory of probability is of interest not only for its own sake, but also for the numerous applications to questions of great importance for society. Then he went on, for about two hours, to set out the basics of the theory of probability. Toward the end of the lecture he turned to one of the promised applications: voting and elections.

Given the variety of opinions among an assembly's members, it is difficult to know, or even to define, its will, he stated and then delved into an attack on the previously accepted method of deciding an issue by majority opinion. Laplace was in total agreement with Borda and Condorcet in his assessment that decisions made by majorities may be defective. If a subject is complicated or delicate, or runs counter to accepted knowledge, it is quite possible, he pointed out, that the truth might lie with the minority. In fact, the larger the assembly, the greater are the chances that the majority decision is wrong. Scientists, he reminds his listeners, can tell many a story about first impressions that turned out to be erroneous. Whatever *seems* true may not always *be* true.

Yet a few rules can render majority opinions much more reliable. If the assembly consists of instructed, well-informed men, guided by common sense, their assessment about an issue has a greater chance of being true. For example, if a hundred men say that the sun will rise the next day, it is very probable that they are correct. Hence, the first rule of ballot decisions is that judges or electors be well informed about the issues about which they are to decide. Therefore assemblies should only be required to render judgment about, or vote on, matters that are within reach of most of their members. From this Laplace deduced the extreme importance of public instruction and the requirement that national representatives be honest and well informed about the subjects at hand. Judges and electors must be guided by truth, justice, and humanity, prerequisites of society that are as necessary to social order as the law of gravitation is indispensable for the existence of physical order.

So what should be done? After explaining the intricacies of the Borda count, Laplace goes on to say that it would, in fact, be the best election method if every voter would rank the candidates *in the order of merit that he attributes to them*. Ignoring the bothersome matter of ties—as in the Grappa, Limoncello, Amaretto example of the previous chapter— Laplace maintains that the Borda count would work well if everyone were completely truthful in his rankings. But passions and private interests abound, he remarks, and considerations that have nothing to do with a candidate's merit may trouble the order in which electors place them. Specifically, electors may put a candidate at the end of the list, not because he is unworthy of a better ranking but because he is a threat to one's preferred candidate.

This needs some elaboration. In Laplace's words, the Borda method might be frustrated by electors who would place the strongest opponents of their favorite candidates at the bottom of their list. Thus supporters of one candidate could deny a strong contender the m-units he would rightly deserve—just so he does not become a threat to their preferred candidate. In our example above, Tom's backers who fear that Dick may win the election would place him last, even if they believe—in their heart of hearts—that Dick is preferable to Harry. Harry would thereby gain 30 m-units for a total of 144, which does not help him at all. Dick however would drop from a winning score of 190 to 160. And Tom, whose score would remain unchanged at 182, would now get the nod. This is what happened when Hardy's supporters placed the unloved Goofy in front of Laurel.

Thus, the gist of strategic voting is: try to wedge inferior people who have no chance of winning between your preferred candidate and his most dangerous contenders. Strategic voting undermines Borda's election procedure because it denies good candidates the m-units they deserve. The upshot is that mediocre and inferior candidates are raised to second or third ranks instead of being banished to the bottom of the list where they belong. This gives a great advantage to candidates of mediocre merit, Laplace tells the students. While getting few top places they would also get few lowest places. Hence, if a bunch of electors decide in concert to vote strategically they may be able to catapult their preferred candidate to the top of the list. But they should beware. If too many voters—who support different candidates—practice the method, strategic voting can backfire. Without actually meaning to do so, an assembly may get Goofy elected. And that is where the real danger lies. Examples of possible mishaps that resulted because some of the electors were too clever by half can be witnessed even today, Laplace contends. He concludes his remarks with the observation that bad experiences with the Borda count had led most establishments who had given it a try to abandon it.

Laplace's criticism had been anticipated by Condorcet. In his *"Essai sur l'application de l'analyse à la probabilité des décisions rendues à la pluralité des voix"* (Essay on the application of probability analysis to majority decisions), published in 1785, he wrote that schemers could cause the method to fail. If two groups support opposing candidates and practice strategic voting, the winner may, by default, be one of the less desirable candidates. Actually, the possible existence of intriguing schemers was the reason why Condorcet proposed his own method. In contrast to Borda's, he claimed, his method would necessarily lead to the election of one of the preferred candidates. Well, "necessarily" is certainly exaggerated. If someone is elected, he will be one of the good candidates. But the emphasis is on the word "if." Because we know, of course, that a Condorcet winner may not exist.

One may think that the possibility of crowning an inferior candidate by mistake would, in itself, serve as a deterrent against strategic voting. However, such a mishap could occur only through the concerted effort of a very large group of voters, which would be difficult to organize and enforce. Hence, an erroneous election of someone like Goofy would be quite rare in practice. But Laplace, as a mathematician, cannot dismiss the possibility, which is why he rejects Borda's method.

To be fair to Borda, it must be pointed out that he had been aware, at least indirectly, of the problem that strategic voting entails. His calculations as to how many votes a candidate would need, assuming a worst-case scenario, in order to win the election are a case in point. The worst-case scenario occurs precisely when electors vote strategically, that is, when supporters of a candidate, say Peter, put his closest competitor, say Mary, second, but all other electors put Mary first and Peter last. Borda provides a mathematical proof that if there are n candidates and Peter receives $1 - 1/n$ parts of the votes, strategic voting cannot hurt him. (See chapter 5, especially the appendix.) In other words, if this stringent requirement is satisfied—with ten candidates it means receiving the support of at least 90 percent of the electorate—the Borda winner and the Condorcet winner are identical. But for anything less the problem of strategic voting remains.

What did Borda have to say in defense of his method? When confronted with the criticism, he retorted that his method was only meant for honest men. With this he meant that honest electors would rank the candidates according to their true merit, without any strategic considerations. As excuses go, this was a rather lame one. If everybody were honest, many of the world's problems would disappear and there would be little need for safeguards.

And what remedy did Laplace suggest? In his lecture before the budding teachers he proposed none. But later he did offer a constructive suggestion. It was put in writing only in 1812, but he must have made it earlier. Laplace advocated the traditional majority vote . . . with a twist. An issue would not be decided, and a leader would not be elected by a simple majority; whoever or whatever gets the most votes would not automatically win. Rather, in order to declare a winner, an *absolute* majority would be required; the candidate would have to garner at least half the votes plus one.

This method has the obvious advantage, Laplace pointed out, that a candidate who is rejected by a majority of electors cannot be elected. So far so good. And if only two candidates present themselves for election, the winner according to Laplace's absolute majority requirement would be identical to the Condorcet winner and to the winner by the Borda count. So far even better. But what if there are three or more candidates? In the above example, neither Tom, nor Dick, nor Harry received the absolute majority of forty-one votes.

So where do we stand? Borda's rank counting method is prone to strategic voting, Condorcet's two-by-two showdowns may produce cycles, and Laplace's absolute majority election may also end without a victory. We seem to have come full circle.

Since no satisfactory election method was known in 1796, the Académie des Sciences adopted the Borda count for the election of new members in spite of its shortcomings. But the honeymoon only lasted until Napoleon made his displeasure known. And suddenly it turned out that some of the students at ENS had paid attention to Laplace's lectures after all; in 1804 the academy replaced Borda's method with the requirement of an absolute majority. From then on, at least half the members had to approve the admittance of a new member. What if no candidate managed to win that many votes? Well then the vacant place would remain unfilled until at least 50 percent of the members plus one agreed on a new candidate.

Laplace's requirement of an absolute majority was all right for the Académie des Sciences, but to leave a country without a leader until more than half the electors agree on a candidate was not. In principle, the requirement of an absolute majority means that elections would have to be repeated until half of the electors plus one settled on a candidate. Laplace realized that this could lead to a never-ending sequence of votes. But suddenly the rigorous mathematician exhibited a more pragmatic side. Experience shows, he wrote, that the general wish among the voters to get the election over and done with, will soon lead an absolute majority of the electorate to settle on a single candidate. Coming from a scholar, committed to strict adherence to the rules of mathematical inquiry, such a statement seems lame. But it is refreshing to realize that even a man of Laplace's stature did not find it beneath himself to bend the rigorous rules of mathematics somewhat when needed.

Nowadays, France's parliamentary and presidential elections take place according to Laplace's suggestion. In order to win, a candidate must garner an absolute majority among the electorate. If there is no Laplace winner in the first round, another round is held two weeks later. This runoff is held only between the two candidates who led in the first round, which is not quite what Laplace had in mind, but—in the spirit of practicality—does speed things up. In the second round, one of the two candidates is guaranteed to receive more than half the votes. The victorious candidate can then fancy himself as the winner according to the majority requirement, the Borda count, the Condorcet method, and Laplace's criterion. And less than half of the voters rejected the new leader. Forget the fact that in the second round he only had to run against one other contender.

Having dealt with voting and elections, Laplace turns to a discussion of criminal trials by judges and juries. Using the tools of probability, he shows that if a majority of only one jury member is required to convict an accused, his culpability is somewhat in doubt since the verdict could have been arrived at by coincidence. Yet demanding unanimity among the jury members would guarantee, with a large probability, that guilty verdicts are just. But there is a problem with that too. Such a stringent requirement would often result in failures to convict. Many truly guilty people would go free and remain dangers to society because the jury could not bring itself to render a unanimous verdict.

Laplace recommends a compromise. If society wants guilty verdicts to be pronounced unanimously, a limit should be placed on the size of the jury. (After all, getting, say, thirty-one judges to agree on a guilty verdict is difficult.) If society prefers a large number of judges, the requirement of unanimity should be abandoned, but a majority larger than one should be required to balance the presumption of innocence with the danger of letting criminals go scot-free. Based on probability calculations, Laplace suggests a majority of nine judges out of twelve in order to condemn an accused, instead of the five out of eight that was customary at the time. Today jury trials in the United States demand unanimity, in England a ten to two majority is required, in Scotland a simple majority of the fifteen-member jury suffices for a guilty verdict.

Before we turn to the next chapter let it be known that, yes, there is, of course a rue Laplace in Paris's 5th arrondissement and yes, there also are extraterrestrial addresses that bear his name. There is a lunar feature, the Promontorium (or mountain crest) Laplace, and Asteroid 4628 is also named after him.

BIOGRAPHICAL APPENDIX

Pierre-Simon de Laplace

When Laplace was thirty-nine, he married the nineteen-year-old Marie-Charlotte de Courty de Romanges. The couple had a boy and a girl. During the Reign of Terror, Laplace fled Paris with his family, moving fifty kilometers outside the capital. He was generally left in peace by the revolutionaries. Only once did they bother him to consult about a new calendar they wanted to institute. It was to start on the autumnal equinox, approximately 22 to 24 of September and have twelve months of three weeks of ten days each. Then five days of holidays would be added at the end of the year. It was easy in terms of arithmetic but, unfortunately, did not quite conform to the astronomical facts. Laplace knew the calendar would also require leap years but he also knew better than to argue with the revolutionaries. Thus, he gave his stamp of approval and the visitors were gone.

After the revolutionaries shut down the Académie des Sciences, the convention authorized the Committee of Public Instruction (the ministry of education) to create a teachers' college, and a year later the École Normale Supérieure was founded. Laplace and the equally famous Joseph-Louis Lagrange were recruited to teach mathematics. Many of today's mathematicians would gladly give an arm and a leg for the possibility of listening to the lectures of these illustrious men. But for the 1,200 budding schoolteachers the pace was too demanding and after only a few months the École temporarily closed its doors again. Napoleon reopened it in 1808, and today ENS is one of the world's foremost institutes of higher learning. After grueling competitions of written and oral examinations only a few handfuls of students are accepted each year; after graduation, they mostly go on to become leading professors and researchers around the world.

Laplace was not too sad to lose his job at ENS because the following year his beloved Académie des Sciences opened its doors again. In addition to resuming his duties there, he was named director both of the Paris Observatory and of the Bureau des Longitudes. His performance in these two positions was mixed at best, with some colleagues complaining that he had little inclination for practical work, pursuing his theoretical interests instead. For a few short weeks, he also served as Minister of the Interior, but the nomination was quickly withdrawn when it became apparent that he was not up to the job. Apparently his mathematical training and quest for rigor got in his way. It was no less than Napoleon, his former pupil, who recognized Laplace's administrative deficiencies. "He sought only subtleties, only conceived of problems and carried the ideas of the 'infinitely small' even into administrative matters." Nevertheless, when monarchy was restored, Laplace lobbied for, and received, the title of Marquis, much to the contempt of many of his colleagues. He died in 1827.

THE OXFORD DON

The theory of voting and elections was not in a satisfactory state at the beginning of the nineteenth century. Majority votes fail to take into account the electors' preferences beyond their top choice, and when the lesser choices are taken into account, cycles appear. The Borda count could result in the election of a candidate whom nobody really wants, while a Condorcet winner, who would beat every competitor in two-by-two showdowns, is not always guaranteed to exist. The theory was in a quandary, but further advances had to wait until an unlikely chap appeared on the scene. He was born as Charles Lutwidge Dodgson but often liked to hide behind a pseudonym. By Latinizing, interchanging, and re-anglicizing his two first names he became Lewis Carroll, author of *Alice's Adventures in Wonderland*. Dodgson was an exceptionally creative person. A prolific writer, a pioneer of photography, and an accomplished mathematician, he is also known as the author of some important papers on voting and elections.

Dodgson's father was a member of the Anglican clergy. He had studied classics and mathematics at Oxford, gaining first-class degrees in both subjects. He became a lecturer in mathematics at Oxford but had to give up the post upon getting married, since members of the faculty were required to remain bachelors. Instead he became a man of the cloth. Reverend Dodgson was the curate of All Saints' Church in Daresbury when Charles was born in 1832, as the third of eleven children. His early education was provided at home by his parents and consisted mainly in the reading of religious texts. But the boy who suffered from a stammer was curious and, wanting to follow in his father's footsteps, took to mathematics. At age twelve he was sent off to a school, ten miles from his home, where he lived in the headmaster's house. He excelled in mathematics and entered the Rugby School three years later.

Upon graduation from Rugby, Dodgson entered Oxford. His studies did not get off to a good start. When he first arrived, in May 1850, no accommodation could be provided for him and he was sent home again. Half a year later, accommodation had been found but two days after his arrival, news reached him that his mother had suddenly died and he again traveled back home. When he finally settled down to study, he did brilliantly in mathematics, gaining a first-class degree, and somewhat less brilliantly in the classics, in which he obtained a third-class degree. His academic achievements sufficed for the award of a scholarship of 25 pounds per year, for life, which was later upgraded into a fellowship. His duties included coaching students and, in 1855, Dodgson became a lecturer in mathematics. Unlike his father, he did not take the vows of a priest because he did not quite believe in all the church's tenets.

As a don, a faculty member of the college of Christ Church, he wrote some treatises about Euclidean geometry and about determinants, but they had no lasting impact on the mathematics profession. Dodgson mainly contributed to mathematics by tutoring students, writing exercise manuals and study guides, and solving puzzles. Actually describing the latter activity as puzzle solving is too condescending. The mathematical questions Dodgson dealt with, often derived from the pages of the *Educational Times*, a remarkable journal that was published monthly, starting in the second half of the nineteenth century. It was devoted to pedagogical matters, with announcements of scholarships, information on vacancies for teachers, book reviews, textbook advertisements, and—most importantly—a section on mathematical problems and their solutions. These problems were of very high quality, and important mathematicians like J. J. Sylvester and G. H. Hardy and the philosopher Bertrand Russell contributed to the pages of the *Educational Times*.

With spare time on his hands, Dodgson decided to take up a hobby. Photography had been invented just twenty years earlier and was still in its infancy when Dodgson ordered his first camera in 1855. He soon became an accomplished photographer, but his choice of subjects leaves a troubling impression: Dodgson had a penchant for taking pictures of young girls. And, as if that had not been a sufficiently questionable activity, as he grew bolder, he preferred photographing them in the nude— albeit with their parents' consent. One of his favorite models was the daughter of the Dean of Christ Church, the noted scholar and coauthor of the famous *Greek Lexicon*, Dr. Henry Liddell. Judging from Dodgson's photographs, Alice Liddell—fully clothed—really was a beautiful eleven-year-old girl.

One day, Dodgson and two colleagues went for a boat ride with the Liddell family. Alice and her sisters got a bit bored, and to while away the time they asked Dodgson to tell them a story. Right then and there, he invented a tale that was so amusing and at the same time so deep that it fascinates children and adults even today. Dodgson was at his best that day, even his stammer disappeared. Later he wrote down everything he had told the sisters and presented the little girl with a written version of the story. A friend of the family caught a glimpse of the manuscript at the Liddell's house and urged Dodgson to publish it. *Alice's Adventures in Wonderland* appeared as a book in 1865 and became a perennial bestseller. Its sequel, *Through the Looking Glass*, followed in 1872 and sold 15,000 copies within seven weeks.

Dodgson's scientific career is rather unremarkable. He did not produce any mathematical results of lasting value. Whatever contributions he made were minor and not terribly important. One of his achievements, for example, was the development of an algorithm—yes, we would call it an algorithm, even though computers were still many years off—to calculate the exact date of Easter Sunday for every year until 2499. By the way, in this he did one better than the German Carl Friedrich Gauss from Göttingen, considered the most important mathematician of the nineteenth century and often called the Prince of Mathematics, who worked out a formula that gave the correct result only until 1999. Dodgson's method worked for all years, with the notable exception of 1954. Why? He tried to find out but resigned after a while and admitted that he "cannot in the least, account for this curious anomaly."

But there was one area to which Dodgson made contributions that endured the test of time. They concerned the theory of voting and elections. His interest in the subject was born out of his concern for the well-being of his college. Not having a family of his own, he kept himself busy by delving into administrative matters of Christ Church. For example, he protested vehemently against converting parks into cricket fields or permitting science students to drop classics . It was a time when Dr. Liddell tried to act out his artistic impulses by commissioning improvements and alterations to the college buildings. Much to the dean's displeasure, Dodgson strongly believed he had to protect the interests of Christ Church by resisting architectural changes. Thus the dean's edificial visions were often frustrated by the don.

Decisions about the students' fate stopped being the privilege of the faculty toward the middle of the nineteenth century. Until then, young men were nominated to studentships by the dean and the canons of the college. As had to be expected, the awards went mainly to the sons of relations and friends. But in 1855 a Royal Commission, of which Henry Liddell had been a member, put an end to this practice. It recommended changes in the administration of the college. Questions of educational policy and about the management of the properties and revenues of the college were put under the control of a Governing Body. It consisted of the dean, the canons, and the students. Moreover, the students were to be a majority in this body. Obviously, members of Christ Church's faculty were not very happy with the curtailment of their prerogatives. When they were told that scholarships were henceforth to be allocated on the basis of academic achievements they blew their top. In a huff, they proclaimed it undesirable to make awards on "mere intellectual merit." Such was Oxford at the time.

The dean's prerogatives were also curtailed, and the new democratic spirit at Christ Church necessitated the making of decisions nearly on a daily basis. No longer could the head of a college decide at the spur of a moment what projects to push and whom to favor. Everything had to be decided by the ballot. A never-ending string of committee meetings ensued and votes had to be taken on all kinds of issues. Finding a fair method to decide between candidates or different courses of action became crucial.

Dodgson wrote his first treatise on voting toward the end of 1873. The occasion was the upcoming election of a Mr. Francis Paget to a Clerical Senior Studentship and the appointment of a Mr. Robert Edward Baynes to the Lee's Readership in Physics. The latter issue was a matter close to Dodgson's heart, since four years earlier he had opposed the establishment of this readership. He considered it an unwarranted intrusion into his sphere of interest. After all, physics—in his view—was nothing more than applied mathematics. He was mollified only after having being assured that the Lee's Reader would concern himself only with experimental findings.

The committee meeting was set for Thursday, December 18. On the Friday preceding the meeting, Dodgson attended an oral examination of candidates and then spent the rest of the day on various chores. In the eve-

ning it occurred to him to investigate the subject of the upcoming decision. It turned out to be much more complicated than he had initially
expected.

With great haste, he set about writing the pamphlet "A Discussion of
the Various Methods of Procedure in Conducting Elections." It took him
only six days to complete. Simultaneously, he explained his ideas to colleagues in the common room. The pamphlet was printed and distributed
to the members of the governing body of Christ Church in a single day,
just prior to the committee meeting. Dodgson, the stammering mathematician, would have been the first to admit that he was no great orator, and
in general his opinion carried little weight among the administrators. But
these were times of upheaval, building activities had strained the coffers
of the college and uncertainty loomed over the future of Christ Church.
The members of the governing body were open to any reasonable suggestion and were prepared to seriously consider even proposals of junior
members of the faculty. They also had the intellectual capabilities to judge
a complex suggestion on its merits.

Dodgson begins his paper with a description of what can go wrong in
the traditional voting schemes. Starting with the majority vote, he remarks
that extraordinary injustices can happen whenever the electors' preferences beyond their top-ranked favorites are taken into account.

In the following example with four candidates and eleven voters, option b would be declared winner in a majority election.

$b > a > c > d$

$b > a > c > d$

$b > a > c > d$

$b > a > c > d$

$b > a > c > d$

$b > a > c > d$

$a > c > d > b$

$a > c > d > b$

$a > c > d > b$

$a > d > c > b$

$a > d > c > b$

Dodgson feels that this totally contradicts common sense because option
a is ranked either first or second by everyone while b—which garners an

absolute majority—is put dead last by five electors. Would it not seem reasonable that a be elected, Dodgson asks?

Then he discusses an example where candidates—or issues to be voted on—are pitted against each other in two-by-two showdowns. The winner of each showdown is then randomly matched against another candidate, until only one candidate is left. Apart from the fact that matchings are random, this is a replay of Llull's third method. Hence the same criticism applies and Dodgson notes that this method gives "preposterous results" since it depends on the order in which the candidates are matched against each other. The method is entirely untrustworthy, he writes, since it turns the election into a mere accident of which candidates are put up first.

Next Dodgson analyzes a multistage scheme. In each round, electors vote for their preferred candidate. The candidate who garners the least votes is eliminated and the process is repeated until only one of them— the winner—is left over. Dodgson again shows by example that this method could result in the exclusion, in the first round, of a candidate who would be most acceptable to all electors. Since this is a novel method I present Dodgson's example in full:

Let us say there are eleven electors with the following preferences:

b > a > d > c
b > a > c > d
b > a > d > c
c > a > b > d
c > a > b > d
c > a > b > d
d > a > c > b
d > a > c > b
d > a > b > c
a > b > d > c
a > c > d > b

Only two electors rank a in first place. Consequently this choice is eliminated immediately. The other choices move up one notch, and the electors' preferences now look like this:

b > d > c
b > c > d

b > d > c
c > b > d
c > b > d
c > b > d
d > c > b
d > c > b
d > b > c
b > d > c
c > d > b

Now it is d who is eliminated since it only gets three votes (against four each for b and c). We have:

b > c
b > c
b > c
c > b
c > b
c > b
c > b
c > b
b > c
b > c
c > b

In this last round c gets six votes against five votes for b. Hence c is the winner. Now compare this with the initial situation where all four candidates were still in the running: eight of the eleven electors indicated that they preferred a to c.

Finally Dodgson suggests the "Method of Marks." In this method each elector has at his disposal a certain number of marks that he can allocate to the candidates. The candidate with the highest total number of marks is declared winner. There is a difference to the m-unit counting method of Jean-Charles de Borda that was described in chapter 5. There the lowest-ranked candidate receives one m-unit, and one additional m-unit is awarded for every additional rank. Dodgson's method allows the elector to allocate *all* of his marks as he sees fit. For example, he may split all his points between the two best candidates, or he may allocate five marks to his preferred candidate, three to the next, and one each to the following three. And therein lies a problem.

This method would be an acceptable method of election, Dodgson remarks, if electors would allocate their marks to the candidates according to their true preferences. But since electors are selfish and devious, he claims, they would most probably assign all the marks at their disposal to the one candidate they prefer, leaving all others with zero marks. Hence, the Method of Marks would coincide with the majority vote, together with all its shortcomings. (As an afterthought Dodgson mentions a method whereby a candidate is proposed, and the electors can either vote for him or against him. The outcome of this method, like all others, could again be counterintuitive. It could happen that a majority of voters would have preferred a candidate different from the one who was elected.)

Doesn't most of the above sound vaguely familiar? It should, because so far the content of Dodgson's pamphlet reflects not much more than the suggestions and criticisms that the Marquis de Condorcet and the Chevalier de Borda had put forward a century earlier. So was Dodgson a plagiarist?

Dodgson can be exonerated. There is evidence that he had read neither Borda's nor Condorcet's writings on the subject of elections. At first blush it may seem difficult to prove a negative—try proving that the Ancient Romans did *not* use wireless telecommunication—but in this case evidence can be proffered. First of all, it is known that Dodgson did not enjoy immersing himself in the works of his predecessors. In fact, apart from general literature he read very little. But let us assume for the moment that he did try to find out what, if anything had already been written on the subject. To do a literature search he would most certainly have perused the library at Christ Church. And in the stalls of this venerable institution can be found the proof that Dodgson had not read the work of his two French predecessors. A copy of *Histoire de l'académie royale des sciences* of 1781 was actually held at the Christ Church library. But the pages that contained Borda's article "*Mémoire sur les élections au scrutin*" were uncut! So by all indications, Dodgson had not read Borda's work.

How about Condorcet's "*Essai sur l'application de l'analyse à la probabilité des décisions rendues à la pluralité des voix*?" This pamphlet was not even available in the library of Christ Church. The closest place where

Dodgson could have found it was the Bodleian Library. This main research library of Oxford University had been founded in 1602 and its collection was much vaster than the one at Christ Church. Indeed, Condorcet's *Essai* was among its holdings. But here too, one of the pages of a section on elections was uncut. Of course, it is conceivable that he read the tractates in some other library, but that is rather unlikely given their availability close to home. So we may accept this as strong evidence that Dodgson had seen neither Borda's nor Condorcet's writings on elections. And of course he had never heard of Ramon Llull or Nikolaus Kues, at least not in the context of elections. So he was no plagiarist.

In the next chapter of his pamphlet, Dodgson introduces an innovation. He maintained that when the different options are proposed, be they candidates or courses of action, "no election" or "no action" should be included as one of the choices. Without the "do nothing" option electors must pick one person from among a bunch of unworthy candidates even though all would have preferred to, say, leave a position unfilled. Or a course of action would be chosen even though a majority of the electors would have preferred "no action" to all presented alternatives.

It is in the third chapter that Dodgson makes his first real contribution. He discusses the Method of Marks and its susceptibility to manipulation by the electors, a practice we called "strategic voting" in earlier chapters. At first Dodgson specifically proposes assigning no marks to the least-preferred candidate, one mark to the next-to-last candidate, two marks to the second-to-last candidate, and so on. If n candidates compete, the top-ranked candidate gets $n - 1$ marks. So far, no advance on the Borda count.

But then Dodgson did present an improvement. He specifically considered the previously neglected situation when two or more candidates are deemed equally attractive. As we will see, by addressing this issue, he simultaneously solved the problem of strategic voting.

Under Dodgson's scheme, the elector groups equally attractive candidates together and places them into a "bracket." How many marks should these candidates be awarded? Well, a strategic voter would try to manipulate the election by putting the preferred candidate in first place, assigning him $n - 1$ marks, and placing all other candidates in a bracket at the end of the list, awarding them zero marks. But here is where Dodgson breaks new ground. His ingenious suggestion is that whenever candidates are put in a bracket, all of them get the same number of marks they would have received had they been the only candidate at that rank. Thus, the top candidate gets $n - 1$ marks and the next gets $n - 2$. Now, if a bracket containing three candidates follows, all three are awarded $n - 3$ marks. Then the number of marks given to the next candidate, or candidates in a bracket, is $n - 4$, and so on. Hence, if the elector puts one candidate at the top and all others into a bracket at the end, the candidates in the lowest bracket would all be given $n - 2$ marks. Thus the purpose of manipulation would be defeated and electors would be induced to avoid low-ranked brackets.

December 18, 1873 arrived and with it came the first test of Dodgson's proposed method. Mr. Paget was elected to the studentship without further ado by traditional methods. But for the readership in physics Dodgson's Method of Marks was utilized. The results of the vote were very close: a certain Mr. Becker received forty-eight marks, Mr. Baynes got forty-seven. (Another candidate landed, far behind, in third place.) Given the novelty of the Method of Marks, it was decided to match the two candidates, who had practically tied, in a traditional direct vote. And that is where the bombshell came. Without the third candidate, Mr. Baynes who had succumbed in the Method of Marks, came out ahead, scoring eleven votes against Becker's nine. Baynes was therefore elected Lee's Reader in Physics. (He served Christ Church in this capacity for nearly half a century, wrote textbooks on heat and thermodynamics, retired in 1919 and died in 1923.)

This was not what Dodgson had had in mind. His method was supposed to avoid paradoxical results. Trying to downplay his disappointment, his diary entry of that day simply noted that the Method of Marks had been used in the committee meeting. It did not mention the fact that the method had failed miserably by not selecting the candidate who was preferred by the majority of committee members. But the odd outcome troubled him and set him thinking.

His chance to improve on his previous proposal came soon. At the same meeting at which the new reader in physics was elected, a committee had been formed to choose a plan and get a cost estimate for the building of Christ Church's belfry. Six architects were asked to submit designs. The committee met several times but there were sharp differences of opinion and the only issue on which the members could agree was that a decision had to be taken. Since the previous Christmas holiday, Dodgson had thought long and hard about how decisions should be made. His ideas on the subject had not yet been fully formed but architectural matters were dear to him and time pressed. He hurried to put down his ideas before the meeting took place. So, half a year after the appearance of his first pamphlet on the subject, Dodgson presented his newest ideas on making decisions in a committee. But they were immature and their publication was premature. He made suggestions as to how to proceed with a vote, without explaining why the new method would function better than traditional methods. And when a problem became apparent, the pamphlet answered with a big question mark.

Dated 13th of June 1874, the pamphlet was titled "Suggestions as to the Best Method of Taking Votes, Where More Than Two Issues Are to Be Voted On." After the embarrassment of having Baynes, the majority's preferred physics reader, turned down by his Method of Marks, Dodgson had the good sense of scuttling it. To his credit, it must be said that he was quite candid about the method's shortcoming. In the preface to the new pamphlet he wrote: "I do not now advocate the [previously proposed] method . . . as a good one to *begin* with. When other means have failed, it may prove useful, but that is not likely to happen often."

What he now proposed was to first check if an absolute majority exists for one of the candidates or one of the courses of action. A piece of paper should be passed around, on which the candidates or the courses of action are listed. The options "elect nobody" or "do nothing" should also be included, of course. Then the electors put their names under their preferred option. If at this stage an absolute majority for one of the candidates or options is obtained, the question can be considered settled and the meeting can break up.

If no candidate achieves an absolute majority, Dodgson suggests that the candidates should be voted on, two at a time. The candidate who wins against all competitors will be considered the absolute winner. We again seem to be covering familiar ground. Dodgson's absolute winner is, of course, the Condorcet winner discussed in chapter 6. Obviously, it was Dodgson's hope that one candidate would turn out to be preferred to all others in two-by-two showdowns. But we know from previous discussions that Condorcet winners do not always exist. Even if one candidate is preferred to most others, there may be another one who beats him, even though he is inferior to all others.

What does Dodgson propose in this case? The answer in one word is: nothing. All he says is that if in the first round no absolute majority has been achieved, and if no absolute winner can be determined in two-by-two showdowns, then one at least knows that opinions are very evenly divided among the committee members. This was not very exciting news and Dodgson realized that his proposal did not measure up to the standards he had set for himself. He closes his pamphlet with the astute observation that "such a state of things is of course very difficult to deal with." And as a summary he sheepishly adds that "the difficulty, though possibly not diminished, will certainly not have been increased by adopting the process I have here suggested." Well said.

On June 18, 1874, Christ Church's Governing Body met to decide on the belfry. It was a stormy session that lasted for five hours. Four of the original six architects had submitted plans and each elector had his favorite. When the plans were voted on simultaneously, Mr. Jackson's tower obtained nine votes, Mr. Deane's arcade five, and Mr. Bodley's gateway two. (A fourth design was struck down before the voting started.) But there were also seven votes in favor of commissioning a new design by Bodley. No proposal had obtained an absolute majority, so the Governing Body resorted to the method proposed by Dodgson. In the two-by-two showdowns Sir Thomas Jackson's tower—which had received most support in the initial poll—was beaten by seventeen votes against, nine in favor of, asking George Frederick Bodley to submit a fresh design.

It is not known what eventually happened to the belfry, but the controversy went beyond its immediate architectural importance. It was the catalyst for Dodgson's further investigation into the subject. In fact, he decided to write a book on the most practicable method of voting. At the same time, a new controversy started to brew, into which Dodgson immersed himself with relish. It concerned the German-born Orientalist and professor of comparative philology and religion, Friedrich Max Müller. This scholar, who had joined Christ Church in 1851, was considered the world's leading expert on Sanskrit and Indian religion and philosophy. His reputation had even spread to the subcontinent, where his writings evoked great enthusiasm. He also took a lively interest in India's political awakening, without ever having visited the country. Müller's texts are required reading even for today's students and researchers. (His research was not universally acclaimed, however. A Roman Catholic bishop considered his lectures as nothing less than "a crusade against divine revelation, against Jesus Christ and Christianity.")

When Müller was about fifty years old, he decided on a change in direction. After twenty-eight years of service to the college, he no longer wanted to teach, believing that this was a task others could do just as well or better. He wanted to devote the rest of his life to translating and editing the sacred books of the East. The university balked at the idea. But when the University of Vienna offered Müller a chair, which would have relieved him of all lecturing duties, the dons at Oxford started having second thoughts. Faced with the prospect of losing a scholar of world renown they put together a proposal according to which Müller would be freed of the obligation to instruct and tutor students, and still retain half his salary. The other half would be paid to a replacement. Dean Liddell was all for it, and the convocation, the legislative assembly that deals with matters of importance to the university, was to make a decision.

The proposal enraged Dodgson. Paying the deputy only half a salary insulted his sense of fairness. And the fact that it was his nemesis, Dean Liddell, who proposed it, made matters worse. He disapproved vehemently of the plan, all the time stressing that it was no feeling of hostility toward his friend Müller that motivated him to do so. The convocation convened on February 15, 1876. Afraid that dissenters would abstain from voting out of fear of being left in a small and therefore conspicuous minority, Dodgson distributed flyers at the entrance of the hall. In true Lewis Carroll fashion he likened the granting of half a salary to Müller, by cutting the new lecturer's salary in half, to a charity event, where a listener was so overcome by the preacher's eloquence, that he transferred to the alms plate all the cash he could find in his neighbor's pocket.

Once the meeting was opened, the discussion quickly got off topic. Completely forgetting that the subject at hand was the pay of the new Sanskrit teacher, speakers outdid themselves in singing the praises of Müller. Dodgson was so exasperated that he saw himself compelled to rise, as the *Times* reported the next day, and ask his colleagues to please stick to the point. On the question of half-pay for a full-time lecturer there were many different opinions, but Dean Liddell's powers of persuasion carried the day. The proposal was approved by ninety-four votes to thirty-five. Müller devoted the rest of his life to the important texts of Hinduism, Buddhism, Taoism, Confucianism, Zoroastrianism, Jainism, and Islam.

The combined work eventually comprised forty-nine volumes and an index, all published by Clarendon Press at Oxford.

Dodgson was upset, not so much about the miserly salary the new philology and Sanskrit lecturer would receive but, rather, that he had lost a showdown with the dean. Even though it was not due to the voting method that he had suffered defeat, Dodgson decided to channel his energy into reexamining the procedures that lead to a vote in a committee. Within a week he had written a third text on the topic, "A Method of Taking Votes on More Than Two Issues." In the words of the twentieth-century Scottish scholar Duncan Black this pamphlet was "the one which entitles [Dodgson] to a position in the theory of elections and committees only a little lower than that of Condorcet."

A year and a half later, in December 1877, Dodgson distributed the booklet to friends and acquaintances. A cover letter asked the recipients for their comments but it is not known whether he received any. In this text he referred to cycles for the first time, albeit without mentioning Condorcet's name, as if their existence was common knowledge. Maybe Condorcet's Paradox was already a well-known phenomenon in England, even though Dodgson was apparently not aware of the Frenchman's contribution.

His paper starts with some commonsense suggestions. If it turns out at a preliminary poll that one issue (a candidate or a course of action) has an absolute majority, the chairman may announce the winner and end the meeting. If there is no absolute majority, then the chairman asks the electors to arrange all options in the order of their preference, and two-by-two showdowns are performed. If one candidate is preferred to each of his competitors by a majority of the voters, this candidate is declared the winner and, again, the meeting ends. So far, Dodgson talks about ground he already covered in "Suggestions as to the Best Method . . ." And his immediate reaction is no better than his previous observation that "such a state of things is of course very difficult to deal with." Dodgson rather lamely suggests that if some issues form a cycle, electors should be given the opportunity for further debate. The hope apparently is that at some point the problem will simply go away; one or more electors, tired of debating, eventually change their minds, the cycle breaks, and an ordering with a clear winner emerges.

It is when filibustering does not produce a winner either, that the party really starts. The main point in Dodgson's third pamphlet, and the one that guaranteed him everlasting fame among devotees of the theory of voting, was his suggestion as to how a cycle should be broken.

Candidates who make up a cycle can, in some sense, be considered equal. None of them is superior to the others because everyone is beaten by at least someone. But Dodgson found that there are candidates who are more equal than others. Let's say electors' votes are such that Alex, Bob, Carl, and Dick form a cycle. (Alex beats Bob, Bob beats Carl, Carl beats Dick, and Dick beats Alex.) If it would only take one elector to change his mind about the relative merits of Bob and Carl in order to make Carl beat Bob, then, Dodgson suggests, Carl should be declared winner. More generally, the candidate who requires the least number of electors to change their opinion should be declared the winner. Let us inspect the following example.

There are eleven electors who must chose among the four candidates Alex, Bob, Carl, and Dick. Their preferences are given in the following columns:

Elector 1:	Alex > Dick > Carl > Bob
Elector 2:	Alex > Dick > Carl > Bob
Elector 3:	Alex > Bob > Dick > Carl
Elector 4:	Alex > Bob > Dick > Carl
Elector 5:	Bob > Carl > Alex > Dick
Elector 6:	Bob > Carl > Alex > Dick
Elector 7:	Bob > Dick > Carl > Alex
Elector 8:	Carl > Bob > Dick > Alex
Elector 9:	Carl > Bob > Dick > Alex
Elector 10:	Carl > Bob > Dick > Alex
Elector 11:	Dick > Carl > Bob > Alex

In a traditional majority election, Alex would receive four votes, Bob and Carl would get three votes each, and Dick one vote. Alex would be declared winner. But seven out of eleven electors prefer Bob to Alex. Furthermore, six electors prefer Carl to Bob, six prefer Dick to Carl, and six prefer Alex to Dick. The complete ranking is Alex > Dick > Carl > Bob > Alex. We have a cycle, and nobody should be declared winner.

But if Elector 11 would change his preferences slightly (by interchanging Carl and Bob) the majority ranking would become Bob > Alex > Dick >

Carl. The cycle would be resolved and Bob would be the undisputed winner. Or if Elector 5 interchanged Bob and Carl, Carl would win. Yet to make Alex or Dick win would require four interchanges. Hence Bob and Carl have more claim to the crown than do Alex or Dick.

Dodgson's proposal is tantamount to finding the candidate who is closest to being a Condorcet winner. The cycle-breaking rule therefore goes as follows: count the number of swaps that each candidate in a cycle would need in order to come out on top, and declare the candidate who requires the fewest swaps the winner. (A swap is defined as an interchange of adjacent candidates in the voters' preference.) The lowest number of swaps that are required to make a candidate a Condorcet winner is called the Dodgson score. If a true Condorcet winner exists, he has a Dodgson score of zero. (He needs no swaps.) Of course, there could be more than one candidate in a cycle with the same, low, Dodgson score. In the above example, both Bob and Carl have a Dodgson score of one. But since this is less than Alex's and Dick's Dodgson-scores the number of potential winners has at least been whittled down from four to two.

But there remains a problem. Determining whether a candidate is a Dodgson winner is no easy task. One needs to find which preferences, of which voters, need to be swapped. Of course, if sufficiently many swaps are allowed, any candidate can be made into a Condorcet winner. But the point is to find the *lowest* number of swaps that is needed to make a candidate come out on top. This problem is hard; in fact, it is very hard as computer scientists found out 113 years after the publication of Dodgson's third pamphlet.

It was in 1989 that three professors of Operations Research—John Bartholdi, Craig Tovey, and Michael Trick—decided to take a close look at Dodgson's suggestion as to how to break voting cycles. They found that the identification of the preferences that must be swapped is not a great problem if only a few electors participate in the election and only a few candidates compete. But as electors and candidates increase, the number of computational steps needed to find the required swaps rises dramatically.

The number of steps that a computer program requires defines its complexity. Specifically, the question is how processing time increases as the size of the input increases. For example, multiplying two one-digit numbers, like 3 and 7, requires one computation. Multiplying two-digit numbers, like 76 and 84, requires five computations: 70 times 80, 70 times 4, 6 times 80, 6 times 4, and then everything needs to be added up. Multiplying two five-digit numbers requires even more computations. So even fast computers take longer when the numbers get bigger. The question is how much longer. As the example shows, the problem of multiplying n-digit numbers grows roughly with the square of n. Problems whose processing time speed grows in proportion to powers of the input size are called P-problems.

Many problems cannot even be handled in P-time. Some of them are said to belong to the class of NP-hard problems. (NP stands for *nondeterministic polynomial* and "hard" stands for ... well, hard.) The time a computer needs to solve such problems grows exponentially with the size of the input. Even with moderately sized inputs, NP-hard problems soon become intractable. Well, Bartholdi, Tovey, and Trick proved that computing the Dodgson score is NP-hard. And that was its death knell.

After publishing the three pamphlets, it was Dodgson's intention to write a book about voting and elections. He never got around to it. In fact, his only other works on the subject were some letters to the *St. James' Gazette* on how to place and rank contestants in lawn tennis tournaments. But he made good use of his knowledge; at dinner parties he routinely asked his friends to arrange the wines in the order of their preferences so as to arrive at a joint decision on which bottles to serve.

In the next chapters I will be talking about a different problem that has been vexing, and still vexes, democracies: the question of allocating seats in parliament.

BIOGRAPHICAL APPENDIX

Charles Lutwidge Dodgson

Dodgson was enamored with Alice Liddell, the eleven-year-old daughter of the dean of Christ Church, and constantly sought her company. His apparent infatuation with the girl has been a source of wonderment—and worse—for nearly a century and a half. One day, toward the end of June 1863, an event occurred that changed everything. Unfortunately, the reader will not learn much about it. Nothing is known about the mysterious occurrence and only speculations exist. The questions surrounding the event have become known as the Liddell-riddle. What is certain is that Dodgson and Alice had been strolling through a wood and that afterward the relationship between them was suddenly and completely broken off. What happened during that fateful walk? Even Dodgson's meticulously kept diaries, thirteen volumes in all, give no clue. Or rather, they give rise to even more speculations: the pages that refer to the dates June 27 to 29, 1863 were ripped out of the diary. Like so much else that remains mysterious, it is also unknown who had done this, and why. It is unlikely that Dodgson himself excised the pages because when he died of pneumonia in January 1898, it was rather sudden and unexpected. Most probably some member of his family wanted to keep the world from learning a terrible secret.

Speculations about what occurred during the walk range from a proposal of marriage that the thirty-one-year-old man made to the eleven-year-old maiden, to outright pedophilia. Actually a marriage proposal to a prepubescent girl may not have been as outrageous in Victorian times as it sounds today, since twelve was the legal age to marry. If the proposal per se was not all that outrageous then maybe the reason for the secretiveness was the fact that Dodgson was rejected as a suitor by the dean and his wife due to his unimpressive standing as a junior mathematician.

The lecturer really did not represent the suitor the Liddels had in mind for their daughter. After all, Oxford was teeming with eligible young bachelors like, for example, Leopold, the son of Queen Victoria and her husband, Prince Albert of Saxe-Coburg and Gotha. Prince Leopold had come to Christ Church to get an education. Even though today most people have no clue as to who Leopold was, while Lewis Carroll is a household name, Dodgson was certainly not considered the prince's equal on the marriage market at that time. Even his first class math degree could not compare with the prince's honorary doctorate in law. The sad fact that he was unfit for marriage with the dean's daughter may have been the reason why the pages were ripped out of the diary. By the way, Alice and Prince Leopold did apparently have a dalliance, but the romance came to naught. This time it was the prince's family who objected to the liaison, Alice being the daughter of a commoner. Serves the Liddells right, I daresay.

So when the alleged marriage proposal was made, the angry parents forbade Dodgson any further association with their daughters. The rejected suitor responded in kind. He published vitriolic pamphlets against the Liddell family. But did Dodgson really propose marriage, or worse, to Alice? Theories abound. One of them has it that Dodgson was actually after Alice's nanny. He dismissed such rumors with an entry into his diary to the effect that Miss Prickett was an unattractive woman. But over a century later, in the mid-1990s, there was a break in the Liddell-riddle. A film director, researching a screenplay for a movie about Lewis Carroll, found a scrap of paper hidden among the records in an archive. It appeared to be an account of the content of the ripped-out diary pages and seemed to indicate that it was not Alice whom Dodgson was courting, but her elder sister Lorina. She was fourteen at the time and highly developed for her age. Nowadays fourteen would not even be considered marginally more respectable than eleven, but in Victorian times Lorina would have no longer been counted as a girl but as a young woman. Whomever Dodgson courted, it did not matter one bit to Mrs. Liddell, who had different plans for all of her daughters. She forbade any liaison and that may have been what Dodgson had entered into his diary.

There are those who reject the theory of the repressed pedophile in favor of the view of Dodgson as a dangerous womanizer. According to this opinion, his alleged penchant toward little girls was only a diversionary maneuver. For example, there are claims that he was romantically involved with—horror of horrors—the dean's wife. More benign assessments of Dodgson's psyche have it that he never really grew up, remaining a big child throughout his life.

Alice ended up marrying Reginald Hargreaves, a young man who was better known as a cricket player than as a student. The wedding took place at Westminster Abbey and was quite a society event. The couple had three sons, the middle one of which was named Leopold. (Prince Leopold, in turn, who was to become the Duke of Albany, named his daughter Alice.) Two of the sons were killed in action during the First World War. When Alice and her surviving son hit on hard times, she put Dodgson's manuscript up for sale with Sotheby's, hoping to raise 4,000 pounds. It fetched 15,400, an unbelievably high price at the time. In 1932, the centenary of Dodgson's birth, the eighty-year-old Alice received an honorary doctorate in literature from Columbia University. She died two years later.

CHAPTER NINE
THE FOUNDING FATHERS

With this we leave the irksome subject of voting and elections for a while to consider a different field of mathematical conundrums that plague democracies throughout the world. It is the problem of allocating seats in a parliament. Everybody would like to assign the number of delegates that a geographical region or a political party sends to the legislature in a fair and equitable manner. Unfortunately we shall see that the questions that will be raised are quite a bit as annoying, perplexing, and sometimes counterintuitive as the problems and paradoxes that occur when voting for a proposal or electing a leader.

* * *

"The House of Representatives shall be composed of members chosen every second year by the people. . . . The number of representatives shall not exceed one for every thirty thousand . . ." Such says the Constitution of the United States of September 17, 1787. The Founding Fathers decided to allocate the seats of Congress to the member states by dividing the number of each state's voters by a number that is at least as large as 30,000. The Constitution also specified that "each State shall have at least one representative."

Other countries chose different allocation methods. The Swiss constitution, for example, is more specific about its legislature. Article 149 reads as follows: "The National Assembly consists of 200 representatives of the people. The seats are allocated according to the population of the cantons." The alpine republic's statements are more specific, but—as we shall see—they are also more problematic.

A confederation consisting of states of various sizes must allocate a different number of delegates to each of them. Of course each state or canton would like to have as big a say as possible and therefore wants to have a large delegation. A method must be found to allocate the seats in the legislature in a fair and transparent manner, without giving rise to arguments.

The American Constitution's requirements of at least 30,000 citizens per representative and of at least one representative for each state was meant to guarantee that not too much power was given to large member states. Thus a small state with 15,000 inhabitants would be entitled to one representative, while a state with ten times as many inhabitants would get five representatives at most, and maybe less.

The Constitution's wording leaves much leeway, however. Since the total number of seats is not specified, it allows for a wide range of numbers of lawmakers. With a population of 280 million, the House would consist of 5,600 congressmen if the divisor were 50,000 citizens, or of 560 if the divisor were 500,000. Both numbers are constitutionally admissible and no amendment would be required . . . except to the Capitol building.

Switzerland, a federation originally of three cantons whose number grew to twenty-five over the centuries, is famous for the precision of its watches and was equally precise when it came to their parliament. The Swiss constitution specified that in each canton's delegation to parliament there should be one representative per 20,000 inhabitants—neither more nor less. With a about 2.2 million Swiss citizens, the National Assembly started out with 111 representatives in 1848, and successively grew, in parallel with the increase in population, to 198 seats in 1928. In order to keep the number of representatives from rising any further, the number of constituents that were needed to send one representative to the capital was increased to 22,000 in 1931 and to 24,000 in 1950. But then the orderly Swiss had enough of adding and removing chairs to and from the chamber every few years, and in 1962 they fixed the number of representatives at 200.

The greater precision of the Swiss constitution did not put order in the house. To the contrary, it created problems. First, there is its wording: the constitution states facts where one would have hoped for a method of seat allocation. Second, and more importantly, the alleged facts are no facts at all. They are incompatible. Allocating a fixed number of seats, be it 200 or any other number, exactly according to the populations of the cantons is impossible for a very simple reason. Since a canton's parliamentary delegation cannot include 3.7 or 16.2 members, 200 representatives can never be allocated exactly according to the cantons' populations. And herein lies the problem that has plagued the United States, Switzerland, and many other countries for the past few centuries. Any allocation that is exactly proportional to the populations contains fractional parts. These remainders must also be allocated. But how?

An instinctive reaction of almost everybody would be to round the seats to the nearest integer. But this does not work at all, as the following example readily shows.

TABLE 9.1
Three-state union with 1,000 citizens, 100 seats to be allocated

State	Population	Percentage	"Raw" seats	Rounded seats
Louisibama	506	50.6	50.6	51
Calyoming	307	30.7	30.7	31
Tennemont	187	18.7	18.7	19
Total	1,000	100.0	100.0	101

There are only 100 seats in the chamber, but the rounding method allocated 101 representatives to the three states.

But first things first. Since the American Constitution was so poor on details, dissent soon arose. The Founding Fathers wanted to create as large a legislature as possible, in order to decrease to a minimum the danger of corruption. (After all, it is easier to corrupt a small group of people than a large one.) According to the census of 1790, and after Vermont and Kentucky had been admitted to the Union, the population of the United States stood at 3,615,920. Using the Constitution's magic number 30,000 as a divisor, that would result in a Congress with 120 seats. Alexander Hamilton, the first secretary of the treasury, suggested a two-step procedure to allocate the congressmen to the individual states: at first, each state gets the number of seats rounded down. The seats left over would then be distributed according to the largest remainders.

Doing the arithmetic and rounding the numbers down, 112 seats were allocated in the first step. Accordingly, Congress presented a bill on March 26, 1792, that would give the eight states who had the largest fractional remainders one additional seat each. For example, Connecticut, which had 236,841 voters, would arithmetically have been due 7.895 seats (236,841 divided by 30,000). The state would have received seven seats in the first round, and since it belonged to the eight states with the largest fractional remainders, it would have received an additional seat in the second round, for a grand total of eight.

Before signing the bill into law, George Washington conferred with his

TABLE 9.2

Census of 1790. 120 seats to be allocated, 30,000 citizens per representative

State	Population	"Raw"	Initial	Final
			Seats allocated	
Connecticut*	236,841	7.895	7	8
Delaware*	55,540	1.851	1	2
Georgia	70,835	2.361	2	2
Kentucky	68,705	2.290	2	2
Maryland	278,514	9.284	9	9
Massachusetts*	475,327	15.844	15	16
New Hampshire*	141,822	4.727	4	5
New Jersey*	179,570	5.986	5	6
New York	331,589	11.053	11	11
North Carolina*	353,523	11.784	11	12
Pennsylvania	432,879	14.419	14	14
Rhode Island	68,446	2.282	2	2
South Carolina*	206,236	6.875	6	7
Vermont*	85,533	2.851	2	3
Virginia	630,560	21.019	21	21
Total	3,615,920	120.531	112	120

*States receiving an additional seat after the initial allocation.

closest advisors. One of them was Thomas Jefferson, whom we already met in chapter 6 when, as ambassador to France, he was a frequent guest at Sophie de Condorcet's salon in Paris. Jefferson was the author of the Declaration of Independence, Washington's secretary of state, and would later become the United States' third president.

Jefferson did not like the bill one bit. He hailed from Virginia, a state whose delegation, though by far the largest, would not be granted any rounding up. In his negative opinion of the bill he was joined by his fellow Virginian Edmund Randolph, the attorney general. On the other side of the argument stood Alexander Hamilton and General Henry Knox, the secretary of war. They argued for adoption of the bill. Knox came from a state that was a candidate for rounding up: Massachusetts with a fractional remainder of 0.844. The only player who seemed to have been quite selfless was Hamilton, whose home state, New York, stood to lose 0.053 seats by the system he advocated.

The naysayers did have a legitimate constitutional argument. As the astute Randolph pointed out, the bill would award all states whose delegations were to be rounded up more than one representative per 30,000 citizens. This would be unconstitutional. For example, New Hampshire would have received one representative per 28,364 citizens (141,822 divided by 5). Washington was still unsure. It was April 4. In two days the bill would automatically become law, even without his signature.

The morning of April 5 arrived. It was the last day a veto could be cast, and Washington had to make a decision. He called Jefferson to his office, before the secretary of state had even had a chance to eat his breakfast. The president was upset. Opinions seemed to be divided not because the apportionment of seats was problematic but because Northern states were pitched against the Southern states. The president did not want to take sides. Jefferson, secretly rejoicing, put Washington's mind at ease. A letter to Congress was drafted in which the president announced his veto against the apportionment bill. "I have maturely considered the Act passed by the two Houses . . . and I return it . . ." he wrote in his message to Congress. One of the reasons he gives for his veto is that "the bill has allotted to eight of the States more than one [seat] for thirty thousand [voters]." It was the first veto in the history of the United States, and only one of two that George Washington would ever cast. Be it noted, by the way, that George Washington's home state was Virginia.

So back to square one it was. On April 10, Congress threw out the vetoed bill and adopted a method of apportionment suggested by Thomas Jefferson. It consisted, first, of determining the preferred size of the House and, second, of finding a ratio that—when results are rounded down—gives exactly the required House size. Thus, the trick was to adjust the divisor according to the size of the House. As we saw above, using Jefferson's method with a divisor of 30,000 would have resulted in a House with 112 representatives. In order to obtain a House with 120 seats, a divisor of about 28,500 would be needed. (Actually, any divisor between 28,356 and 28,511 would have worked.) But this number leaves a very bad feeling. Even though, on average, it satisfies the constitutional requirement of one congressman for at least 30,000 voters (3.6 million voters for 120 seats) for the United States as a whole, it violates the requirement for some states. After all, it was the unconstitutionality of a divisor lower than 30,000 for each individual state that had prompted Washington to cast his veto. To circumvent problems, Congress decided that the House would comprise 105 members. With the number of seats set, it was determined that one representative per 33,000 citizens would do the trick. Now everything seemed in order.

Jefferson's method is referred to as a "divisor method." It remained in force for fifty years, until 1830. In the meantime the Union grew from fifteen states to twenty-four and population increased to nearly 12 million. To accommodate the increasing population of the growing Union, the number of seats in the House grew from 104 to 240. But the principle remained the same: decide on the size of the House, find an appropriate divisor and round down.

Not everyone was happy with the system, however. The small states started getting restive. They noticed that something was amiss. The big brothers, like Virginia, always seemed to be getting more than their fair share. It soon became apparent that Jefferson's method—which had given Virginia an advantage in 1790 but otherwise seemed fair enough—put small states at a disadvantage. Delaware, for example, got rounded down four times, with "raw" numbers of seats of 1.61, 1.78, 1.68, and 1.52. The state of New York, with quotas of 9.63, 16.66, 26.20, 32.50, and 38.59, got rounded up every single time.

One reason lay in the fact that rounding 3.5 representatives down to 3 hurts more than rounding 30.5 down to 30. Thus a small state would require more citizens for each representative.

TABLE 9.3
Jefferson's round-DOWN method

Union of two states, total population 340,000, 33 seats in the House of Representatives. The ratio of citizens per representative (the "divisor") that is needed to achieve this House size, using the round-down method, has been determined as 10,000.

State	Population	"Raw" seats	Seats	Ratio
Massaware	305,000	30.5	30	10.167
Louisylvania	35,000	3.5	3	11,667
Total	340,000	34.0	33	10,303

The smaller state requires about 15 percent more citizens for each representative than the larger state (11,667 versus 10,167).

There is another reason why small states get booted most of the time. It is mathematical and a bit subtler. Let me first give a numerical example. (Note that "raw" seats of 26.20 and 1.68 in table 9.4 correspond to a true situation in New York and Delaware, mentioned above.)

TABLE 9.4
Jefferson's round-DOWN method

With a population of 10 million and 100 seats in the House, the initial divisor is 100,000. Twenty-eight seats are to be allocated to Neware and Delayork. In order to apply Jefferson's method, the divisor is reduced from 100,000 to 97,000.

	Population	Divisor 100,000	Seats	Divisor 97,000	Seats	Ratio
Neware	2,620,000	26.20	26	27.01	27	97,037
Delayork	168,000	1.68	1	1.73	1	168,000
.		.		.		
.		.	72	.	72	
.		.		.		
.						
Total	10,000,000		99		100	100,000

Delayork needs 73% more citizens per seat in the House.

"Neware" gets the additional seat, in spite of initially having had the lower fractional remainder (0.20 versus 0.68). The explanation for this manifestly unsatisfactory situation is that when the divisor is reduced from 100,000 to 97,000, a smaller population is needed for every seat already assigned. In the above case, each of Neware's 26 initial seats requires 3,000 fewer citizens, thus granting this large state a 27th seat. At the same time the small state of "Delayork" profits only once from the lower divisor. Another way to see this is to realize that the lower divisor increases the number of raw seats by 3.1 percent. Thus Delayork's raw seats increase from 1.68 to 1.73, while Neware's seats increase from 26.20 to 27.01, putting it a hair's breadth beyond the threshold for an additional seat. The upshot of all this is that while Neware needs less than 100,000 citizens per congressional seat, Delayork requires a whopping 168,000.

Not surprisingly, all this is patently unfair to the small states, and they finally caught on. They complained that rounding down left some of their voters unrepresented. They certainly had a valid point. Voters who make up the fractional parts of a seat effectively got rounded out of the system. The small states found an advocate for their case in the person of John Quincy Adams. This former president and elder statesman, whose home state, Massachusetts, was the second largest in the Union, no longer had any ax to grind, neither for himself nor for his state. Deeply troubled by the fact that Jefferson's method effectively disenfranchised many voters, he became a spokesman for the small states. After passing many a sleepless night, he announced that he had found a system that would end the discrimination of small states.

Adams had not looked very far for a remedy to this ill of society. In fact, he decided to advocate Jefferson's method with only a teeny difference: after performing the initial computations, the number of seats would be rounded up instead of down. In his view, this corresponded more closely to the spirit of the Constitution, since by rounding the fractional parts of a seat up to a full seat, every citizen would be represented, and then some. Of course, the method would give the small states an advantage. But this rounder-upper apparently felt that a little affirmative action would do no harm after so many years of discrimination.

TABLE 9.5
Adams's round-UP method

With a population of 10 million and 100 seats in the House, the initial divisor is 100,000. 28 seats are to be allocated to Neware and Delayork. In order to apply Adams's method, the divisor is increased from 100,000 to 104,000.

	Population	Divisor 100,000	Seats	Divisor 104,000	Seats	Ratio
Neware	2,668,000	26.68	27	25.65	26	102,615
Delayork	120,000	1.20	2	1.15	2	60,000
.		.		.		
.		. }	72	. }	72	
.		.		.		
.		.		.		
Total	10,000,000		101		100	100,000

In spite of initially having the lower fractional remainder (0.20 versus 0.68) Delayork gets the additional seat. Neware needs 71% more citizens per seat in the House.

The shoe is now definitely on the other foot. With an increase of the divisor from 100,000 to 104,000, a larger population is required for every seat. Neware would have needed 4,000 additional citizens for each of the 27 seats. Since it does not have them, its delegation obtains only 26 seats. Delayork has citizens left over and gets rounded up to 2 seats. In the final count, Delayork needs only 60,000 citizens per representative, while huge Neware requires over 100,000.

As had to be expected, the large states would have none of that. Affirmative action was not their forte and—being the stronger side in this dispute—the rounder-downers got their way. Adams's suggestion, which is sometimes called "the method of smallest divisors," was considered by

Congress but never enacted. "I hung my harp upon my willows," Adams wrote in his memoirs, and simply gave up.

It took the rhetorical skills of Senator Daniel Webster, one of the most eloquent Americans to have ever walked the floor of the Senate, to convince Congress to adopt the course of action that reasonable people would have found most sensible had they not been so caught up in looking out for themselves. Webster, a lawyer by profession, was first catapulted to prominence by his famous defense of Dartmouth College's independence against the New Hampshire legislature. He was a spellbinding orator and his speeches are considered even today rare examples of rhetorical brilliance. Whenever he addressed the Senate, it was standing room only. Men and women traveled from afar to hear him speak, and the moment he took to the podium, all present fell silent. Reportedly, his perorations could move the most reserved of men to tears.

With such credits, it is somewhat surprising that the allocation method Webster came up with was very simple indeed. Again, it was the procedure that Jefferson had originally proposed, but this time with a two-sided twist. It consisted of finding a divisor for the populations of the states, such that the result, when rounded *up* or *down* to the nearest integer number, gave the desired amount of seats. The "method of major fractions," as it was thereafter called, would not avoid inequities but at least it was unbiased; it would favor large states sometimes, small states at other times. In 1842, Congress, which since 1787 comprised the Senate and the House of Representatives, adopted Webster's method.

TABLE 9.6
Webster's round-to-nearest-number method

Union of two states, total population 330,000, 33 seats in the House of Representatives. The ratio of citizens per representative that is needed to achieve this House size has been determined as 10,000.

(1)	State	Population	"Raw" seats	Seats	Ratio
	Coloraska	304,000	30.4	30	10,133
	Nebrado	26,000	2.6	3	8,667
	Total	330,000	33.0	33	
(2)	State	Population	"Raw" seats	Seats	Ratio
	Oregansas	296,000	29.6	30	9,867
	Arkanson	34,000	3.4	3	11,333
	Total	330,000	33.0	33	

Sometimes, as in (1) above, the small state needs fewer citizens per representative, at others the large state profits.

The ever so reasonable method of major fractions stayed in force for not more than ten years. Maybe it was too reasonable a procedure, because soon the squabbles started again. In 1850, before anybody had a chance to get all worked up about the results of this year's census, Senator Samuel Vinton from Ohio stepped in. His objective was to put an end to the bickering and quarrelling that, like clockwork, took place every ten years, after each census. He proposed a new method. Each state would, first of all, be allocated the number of seats rounded down. The leftover seats would then be distributed to the states with the greatest fractional remainders.

Only the proposal wasn't so new. In fact, it was precisely the one that Hamilton had proposed half a century earlier. It was also precisely the one that had been vetoed by George Washington. But a name change avoided a replay of the debacle of 1792, and the method that became known henceforth as the "Vinton method" was enacted by Congress. (Hamilton would have rejoiced at the rehabilitation of his method, had he not been killed in a duel in 1804.) To make everybody happy, the House was also increased from 233 seats to 234, a size in which Hamilton's and Webster's methods actually agreed.

The population of the United States continued to grow, and with it the size of the House was increased again and again. In 1860 the number of congressmen representing their constituents increased to 241 from 234 ten years earlier, and in 1870 the number of seats was going to be fixed at 283, a number in which Hamilton's and Webster's methods again resulted in the same apportionment. But because of political squabbles the House's size was subsequently increased to 292 seats. Now everybody was unhappy, because the final apportionment did not agree with either method.

Then something extraordinary happened. After the results of the census of 1880 became known, everybody expected the House to grow again. In order to give the congressmen the necessary ammunition for the infighting that would undoubtedly precede the next apportionment of the House, C. W. Seaton, the chief clerk of the Census Office, did some computations. Using the census results of 1880, he worked out the apportionments according to Vinton's method for all House sizes between 275 and 350. Starting with 275 representatives everything worked out just fine all the way up to 299. Whenever he added an additional seat it was picked up by some lucky state. But when he reached 300 seats, a bombshell fell in his lap. The delegation of the state of Alabama *decreased* by one represen-tative, from 8 to 7. In its stead, *two* states, Illinois and Texas, each got an additional seat.

Seaton was dumbfounded. The congressmen were flabbergasted. How could such a thing happen? The phenomenon became known as the Ala-bama Paradox.

TABLE 9.7
Alabama Paradox

299 seats are to be allocated. Total population is 49,713,370 and the appropriate divisor is 165,120.

	Alabama	*Texas*	*Illinois*
Population	1,262,505	1,591,749	3,077,871
"Raw" allocation	7.646	9.640	18.640
Seats in first round	7	9	18
Fractional part	0.646	0.640	0.640
Additional seats	1	0	0
Total seats	8	9	18

Now 300 seats are to be allocated. The appropriate divisor is 164,580.

	Alabama	*Texas*	*Illinois*
Population	1,262,505	1,591,749	3,077,871
"Raw" allocation	7.671	9.672	18.701
Seats in first round	7	9	18
Fractional part	0.671	0.672	0.701
Additional seats	0	1	1
Total seats	7	10	19

Alabama loses one seat; Texas and Illinois each gain a seat.

The reason for this paradox becomes apparent when we delve a little deeper into the numbers. When the total number of seats increases from 299 to 300, the states' "raw" numbers of seats grow on average by about one-third of 1 percent. But Texas and Illinois start out with larger popula-tions and therefore gain more in absolute numbers. Thus the number of "raw" seats grows by only 0.025 in Alabama (from 7.646 to 7.671), by 0.032 in Texas (from 9.640 to 9.672), and by 0.061 in Illinois (from 18.640 to 18.701). As a consequence, the larger states creep past Alabama.

Actually, the phenomenon had already been noted ten years earlier. Rhode Island had had two representatives in the House since 1790. But following the census of 1860, the Plantation State, to its consternation, was allowed only one congressman out of 241. Ten years later, with an increase in population and in the size of the House, Rhode Island hoped to get its second representative back. Initial calculations showed that this would, in fact, be the case if the number of congressmen were increased to 270. If the House were enlarged to 280 members, however, Rhode Island stood to lose that second congressman again. So the Alabama Paradox should have correctly been called the Rhode Island Paradox in 1870. But then the powers to be decided to set the size of the House at 292. Rhode Island got its second representative and the matter was forgotten for another ten years.

In 1880, however, Congress went into a tizzy. The Hamilton-Vinton method of apportionment, which had by now become dear to all, was in danger. Tempers ran high. One congressman accused another of "committing a classic rape on a cloud of statistics, right in the face of the House." In order to prevent the contest between the proponents of the two methods from turning even uglier, Congress decided not to decide and resolved, instead, to enlarge the House to 325 seats. With this size the congressmen did not have to take sides because Webster's and Hamilton's methods agreed and the problem could be postponed for at least another ten years. Maybe a wholly different apportionment method would be found in the meantime? Or the methods would again agree? Or the congressmen would no longer be in Congress and could let their successors worry about the Alabama Paradox.

They were right. All it took in 1890 was an increase to 356 seats and the same compromise could be forged. With a House of that size, both methods agreed, no state would lose a seat as compared to the previous apportionment. Ten years later, no such luck. When tables on the apportionment were prepared in 1901 for sizes of the House between 350 and 400, Maine's apportionment oscillated between 3 and 4 seats and Colorado would receive 3 seats for every size of the House, except at 357 seats where it would be allocated just 2. Of course, the chairman of the Select Committee on the Twelfth Census, no friend of Colorado's and Maine's, suggested fixing the size of the House precisely at 357. Tempers rose and the atmosphere again became ugly.

And more bad news was on the way. The congressmen and the pencil pushers had failed to notice another, even more serious threat to Hamilton's method of apportionment. The nation's population was continually rising and with it the size of the House was increased. But could problems arise even with a House of constant size?

Yes, they could. Let me illustrate by example. In 1900 the populations of Virginia and Maine stood at 1,854,184 and 694,466 citizens, respectively. During the following year Virginia's population grew by 19,767 citizens (+1.06 percent), while Maine's increased by 4,648 (+0.7 percent). If an additional seat were to be allocated to one of the two states, one would think that it would have to go to Virginia. Far from it. The surprising fact is that by Hamilton's method of allocating leftover seats to the states with the largest remainders, it would have been Maine that would have received an additional seat, while Virginia would have lost one. Let us look at the numbers:

TABLE 9.8
Population Paradox

	1900	Seats		1901	Seats	
	Population	raw	rounded	Population	raw	rounded
Virginia	1,854,184	9.599*	10	1,873,951	9.509	9
Maine	694,466	3.595	3	699,114	3.548*	4
Total	74,562,608		386	76,069,522		386

*rounded up

Total population grew from 74,562,608 to 76,069,522 and the appropriate divisors were 193,167 in 1900 and 197,071 in 1901. Virginia's population grew by 19,767, while Maine's grew by only 4,648. Nevertheless, if a new House had been appointed in 1901, Virginia would have lost a seat to Maine. (The numerical values for 1901 are inferred from the population growth between 1900 and 1910. There was no separate census in 1901.)

The reason for this strange situation, which would henceforth be known as the Population Paradox, is that the remainders traded places. In 1900, Maine had the smaller remainder behind the decimal point (0.595) and failed to make the mark. A year later—because the nation as a whole grew faster than either of the two states—it would have been Virginia that would have had the smaller remainder (0.509) and Maine would have picked up the extra seat.

Why is the Population Paradox even more of a threat to Hamilton's method of apportionment than the Alabama Paradox? Well the latter, which appears when the size of the House increases, can be avoided if Congress decides to hold the number of its members constant. But population increase cannot be stopped and so this paradox was here to stay.

This time Congress did take a stand and Hamilton's method was abandoned in favor of Webster's. At least it did not suffer from the defect of the Alabama and the Population paradoxes. In addition, the House was enlarged to 386 seats, which ensured that no state would lose a seat. (Amazingly, it is not absolutely clear whether it really was Webster's method or Hamilton's that was used in 1901. Depending on which population data one looks at, the preliminary data that was used by the House or the final data published by the Census Bureau, one could arrive at either conclusion.)

And this was still not all. Another paradox was looming just around the corner. In 1907, Oklahoma joined the Union. The House consisted of 386 seats. This time the congressmen thought they knew exactly what needed to be done to keep everybody happy. Oklahoma's population stood at about 1 million, which corresponded to five seats in Congress. So the congressmen decided to simply add five seats to the House. Oklahoma would receive them, nobody would get hurt, and everybody would be happy.

Or so they thought. The five seats were added and to nobody's surprise, when the new total of 391 seats was reallocated, using Hamilton's method again, Oklahoma got them all. But something strange happened on the way. New York lost a seat, which Maine picked up! It was quite infuriating. For once everybody had done everything right, and then this happened. The situation was called the New State Paradox. Let us see how it came about.

TABLE 9.9
New State Paradox

	Population before incl. of Oklahoma	Seats		Population after incl. of Oklahoma	Seats	
		raw	rounded		raw	rounded
New York	7,264,183	37.606*	38	7,264,183	37.589	37
Maine	694,466	3.595	3	694,466	3.594*	4
Oklahoma	—	—	—	1,000,000	5.175	5
Total	74,562,608		386	75,562,608		391

*rounded up
After addition of the five seats—which went to Oklahoma—New York lost a seat to Maine.

Even though the populations of New York and Maine did not change, and Oklahoma was awarded exactly the five additional seats, apportionment of the remaining seats was affected. When adding Oklahoma's population to the nation's total, both New York's and Maine's and all other states' fractional seats decreased. But New York, being the largest state, lost more in absolute terms than the smaller states. As a result, Maine's fractional seat managed to inch past New York's remainder.

By the way, the number of electors each state has in the Electoral College corresponds to the number of representatives and senators that the state sends to Congress. Hence the apportionment problem spills over to presidential elections. In 2000, George W. Bush defeated Al Gore by a tally of 271 to 266 in the Electoral College (using the apportionment method that will be described in the next chapter). Had Jefferson's method been used to apportion the House after the 1990 census, Gore would have garnered 271 electoral votes and become the president.

To close this chapter let me summarize the advantages and disadvantages of each of the methods in the following table.

TABLE 9.10

Method	Hamilton	Jefferson	Adams	Webster
Bias toward:	large states	large states	small states	Neither
Paradox				
Alabama	Yes	No	No	No
Population	Yes	No	No	No
New State	Yes	No	No	No

Webster's method seems the most reasonable way of apportioning seats in the House, while Hamilton's, which falls prey to all paradoxes discovered so far, seems the odd man out. But Hamilton's has one "redeeming feature": it favors the large states. So, larger states would continue their battle against Webster and further troubles were unavoidable. (There is one other redeeming feature to Hamilton's method, but we will have to wait until chapter 12 to discover it.)

THE IVY LEAGUERS

Frustrated and dispirited, politicians looking for solutions—not always mathematical, as often as not political—to the seemingly intractable problems of apportioning seats in Congress finally turned to a professional, Walter F. Willcox. Willcox, professor of social science and statistics in the department of philosophy at Cornell University, had been active in the census of 1900 and later became the Census Bureau's chief statistician for population. He was instrumental in raising the debate about apportionment from the lowlands of politics to the realms of science.

The Census Bureau had been established very recently. At the outset, in 1790, the first census had been conducted by U.S. Marshals and for the next one hundred years counting the population was an ad hoc affair. Each time, after the data were published, the office would shut down until the next census. Toward the end of the nineteenth century the need for a permanent agency became apparent. Willcox explained how the establishment of the bureau was brought about: "Director William R. Merriam handled Congress very cleverly; got a stunning group of girls on his staff; nearly all of them, no doubt, wanted to remain in Washington and in the Census Office—at least until they got married. These girls, I was told, brought so much pressure on Congress that . . . the office was made permanent, not for any scientific reason, but to keep the staff from being disbanded." By an act of Congress in 1902 the Census Bureau became a permanent organization.

Willcox was the founder of sociology at Cornell. Born in 1861 in Reading, Massachusetts, he studied at Amherst as an undergraduate and then at Columbia University where he obtained LLD and PhD degrees. He also spent a year at the University of Berlin in Germany. Upon receiving his doctorate, he joined the faculty of Cornell University where his career would span forty years. (More on Willcox and on this chapter's other dramatis personae will be found in the additional reading section.)

The twentieth century's first census in the United States took place in 1910. It revealed a 20 percent population growth as compared to the previous decade. The United States' total population had grown from less than 75 million ten years earlier, to over 91 million. But higher birth rates, immigration, and the addition of Oklahoma do not provide a complete picture. Beyond the increase in the total number of citizens, the distribution of the population within the United States had changed. Poor farmers had begun to look for their fortunes in the cities, and migration from rural states toward urbanized centers resulted. The new realities needed to be reflected in the composition of the House. What method should be used to apportion the seats of Congress? The rural population was not going to take the erosion of their political power lying down. In an unlikely coalition, wealthy landowners and poor farmers were going to fight for their political rights, even if the rights weren't quite on their side. But first things first.

After studying the various apportionment methods, Willcox became convinced that Webster's technique of "major fractions" was the way to go. (Recall that this technique requires finding a divisor for the populations of the states, such that the result, when rounded *up* or *down* to the nearest integer number, gives the desired amount of seats.) It neither succumbs to any of the paradoxes, nor is it biased toward large or small states. Persuaded by the arguments, Congress began to lean toward the Webster-Willcox method. But Ohio, the 4th-largest state, and Mississippi, the 21st, did not like this one bit because they would each have received an additional seat under Hamilton's scheme. (Maine, the 34th state in terms of size, and Idaho, the 45th, would have each lost one.) So, to keep everybody happy, Congress adopted Webster's method but simultaneously increased the number of seats from 386 to 433. (One additional seat each was reserved for Arizona and New Mexico for whenever they would join the Union.) This occurred in 1912 and from then on, until today, the House comprises 435 seats. With this number of seats, Ohio and Mississippi kept the seats that had been allocated to them after the previous census, Maine and Idaho received the additional seat that Webster's method awarded them, and no other state would lose a seat. The fact that the inflation in the number of seats had eroded each representative's voice by about 11 percent apparently went unnoticed ($386/433 - 1 = -11$ percent).

But a sense of unease pervaded Congress. Whatever it was that had made Maine the odd man out, both in the Population Paradox and the New State Paradox, it gave everyone cold feet. What happens to Ohio and

Mississippi, Maine and Idaho, Alabama, New York, or Virginia one day could happen to any other state on a different day. A fresh idea was needed for the next apportionment, due in 1920. Eventually, one did come up. It was the brainchild of Willcox's colleague at the Bureau of the Census, the chief statistician of the Division of Revision and Results, Joseph A. Hill.

Looking for the fairest manner of apportionment, Hill considered the number of constituents that are required for one representative as the key variable. Evenhandedness requires, he felt, that the relative difference in this number between states be minimized. If one state requires 200,000 people to send a representative to Congress, and another state needs only 190,000, then the relative difference is 5 percent. Hill felt that seats should be apportioned in such a way that this relative difference be as small as possible. The following example shows how the apportionment is to be computed.

TABLE 10.1
The Hill method of apportionment

Total population: 4 million. 20 seats to be apportioned. Population per seat = 200,000

| | | | Option 1 | | | Option 2 | | |
| | | "Raw" | Allocated | Citizens/ | | Allocated | Citizen/ | |
State	Population	seats	seats	seat	Diff	seats	seat	Diff
A	3,300,000	16.5	16	206,250	17.8%	17	194,117	20.2%
B	700,000	3.5	4	175,000		3	233,333	
Total	4,000,000	20.0	20			20		

Since the relative difference in Option 1 (17.8%) is lower than in Option 2 (20.2%), the former is preferred.

How would Webster's method of rounding to the nearest integer value work in this example? Well, it wouldn't. In the rare case that the fractional seats of both states are precisely 0.5, Webster's method would give no indication as to which state's delegation to round up or down. The situation would be a tie.

Hill's idea appealed to Edward V. Huntington, a professor of mathematics and mechanics at Harvard University. Actually, Huntington's interest in apportionment was a sideline. The professor had known Hill from his undergraduate days at Harvard where they had been classmates. After studying all apportionment methods, Huntington became an avid advocate of Hill's proposal, which he called the "method of equal proportions," but which would henceforth be known as the Huntington-Hill (H-H) method.

Huntington formalized Hill's argument. An apportionment is good, he stated, if no reallocation of a seat from one state to another can reduce disparity. Of course, this left the notion of disparity to be defined. Huntington interpreted it—as did Hill—as the percentage difference in the number of people that is required for a congressional seat. The allocation of seats that minimizes the disparity is the best apportionment.

To recapitulate, the Webster-Willcox (W-W) method requires finding an appropriate divisor such that the seats—when rounded up or down—add up to 435. (For states that would be allocated less than one seat, the number would be rounded up in any case.) The H-H method requires finding a divisor and then rounding up or down, such that the relative difference in the numbers of constituents is decreased. Like many academic controversies, the dispute became fierce. As more scholars entered both sides of the fray, the proponents of the two methods became known, respectively, as the "Cornell school" (W-W) and the "Harvard school" (H-H).

Which is the preferable method was not only a question of mathematical superiority but also one of which method would be easier to explain to Congress. The Harvard school's method is not an easy one to implement. Not only must several divisors be tried, but it must also be verified for each one whether the reallocation of a seat between two states decreases the relative difference in the number of constituents required for a seat. Fortunately there is an easier way to implement the H-H method.

To explain it requires some preliminaries. Rounding up or down at the midpoint between two integers, that is, at a fraction of 0.5, is called "rounding at the arithmetic mean" because its computation requires just arithmetic. For example, the arithmetic mean is computed by adding one number to another and dividing the result by two (2 plus 18, divided by 2 equals 10). There is another mean, called the geometric mean. It is computed by multiplying the two numbers and then taking the square root. For example, multiplying 2 by 18 and then taking the square root gives 6. Thus the arithmetic mean of 2 and 18 is 10, while the geometric mean is 6.

Huntington claimed that finding an appropriate divisor and rounding at the *geometric* mean gives an apportionment that minimizes the relative difference in the number of people required for a congressional seat. Hence it produces exactly the result that the H-H method seeks.

Thus, rather than rounding exactly halfway between, say, 3 and 4 or 17 and 18, as the W-W method requires, the H-H method demands rounding at 3.4641 (square root of 3 times 4) or at 17.4928 (square root of 17 times 18). The difference between arithmetic and geometric means is quite small for consecutive integers, but for apportionment purposes it may mean the difference between getting an additional seat and not getting it.

Rounding at the geometric mean is a simple way of implementing the H-H method. But is it legitimate? Why should rounding at the geometric mean be related to minimizing relative differences in the number of people needed for a congressional seat? The astonishing statement that both procedures give identical results requires proof. Huntington provided it in an article in the *Transactions of the American Mathematical Society*. The paper, based on seven talks he had given to various audiences during the previous eight years, appeared in 1928. (It would lead us too far astray to reproduce the proof here, and we postpone this to the appendix.) Thus he not only formalized Hill's argument and found a simpler way of implementing it, he also gave it a mathematical underpinning.

Unfortunately, the discovery that rounding at the geometric mean is equivalent to the H-H method also brought to light a serious problem. Fair as the method seems at first blush—who would want to argue that relative differences should not be minimized?—it slightly favors smaller states. A small state requires a fractional seat of only 0.4142 to get rounded from one seat to two, whereas a large state would need 0.4959 fractional seats—nearly 20 percent more—to get rounded from 31 to 32 ($\sqrt{[1 \times 2]}$ = 1.4142; $\sqrt{[30 \times 31]}$ = 30.4959).

This is borne out by the facts. As compared to the Cornell (W-W) method, the Harvard (H-H) method would have allotted one additional seat each in 1920 to the relatively small states of Vermont, New Mexico, and Rhode Island and taken away one seat each from the relatively large states of New York, North Carolina, and Virginia. Obviously, the latter did not take kindly to H-H. And the less populated rural states did not take kindly to the W-W method; when compared to H-H it would have cost them eleven seats.

The squabbles between the supporters of Cornell and Harvard went on and on. In the end, it was all for naught. Faced with implacable opposition to both proposals, Congress at first attempted to find a compromise. It tried to apportion the current 435 seats according to Webster's time-proven method. No agreement could be reached. Next, Congress considered an increase in the size of the House to 483 which would have, again, guaranteed all states at least as many seats as in the previous House—albeit with an erosion of again 11 percent due to the inflation in seats. This bill died in the Senate.

Other attempts fared no better. Congressmen and senators from the rural states blocked all reapportionment legislation and ensured that every attempt to change the status quo was doomed from the outset. Deadlock resulted and in the end Congress once again decided not to decide. In direct violation of the Constitution, no reapportionment was effected in 1921, and the composition of the House remained unchanged from its composition after the census of 1910. The gentlemen from the rural states welcomed this outcome. They were quite happy to postpone the inevitable for another ten years.

But looking the other way was no answer for the long run. With the end of the 1920s and the next census drawing nearer, a decision became more urgent. Congress could not snub the Constitution again; a solution had to be found. The debate reached acrimonious tones with the advocates even stooping to personal attacks. While Hill was a disimpassioned intellectual, Huntington was anything but. Pugnacious, with a taste for verbal sparring, he took up the fight for Hill's and his method with religious fervor.

The fact that the House Committee on the Census did not accept his method of equal proportions was "attributable entirely to one man, Professor W. F. Willcox, of Cornell," he charged in the journal *Science* in February 1928, claiming that this man's "entirely false description ... supported by impressive charts and diagrams" had the effect of completely misleading the congressional committee. Calling Willcox's technique of major fractions "an obsolete method" and an "erroneous idea" he extolled the virtues of his own method of equal proportions, which for its "simplicity, directness, and intelligibility leaves nothing to be desired."

The harangue continued. While Willcox maintained that mathematicians and statisticians were in favor of his method he, in fact, quoted supporting statements only from constitutional lawyers and professors of political economy, Huntington claimed. Moreover, even their backing had been secured by misinformation. He contrasted this with his method of equal proportions that was "endorsed by a general consensus of scientific opinion," being "the only method that has the approval of any organized body of scholars." Omitting to specify who exactly the scholars were, he called upon Willcox to publish his views in some regular journal, not just in testimony before Congress, "so that they may be accessible to the scrutiny of all groups of scholars."

Half a year later, in December 1928, the pages of *Science* again served Huntington for a comparison between his "scientific method of equal proportions" and the "unscientific method of major fractions." Willcox's statements about the latter are "at variance with known mathematical facts" and serious errors in congressional hearings "will be the source of confusion to future students of the problem," he fumed. What was especially galling in his opinion was that "the appearance of such misstatements . . . in a permanent public document gives Congress a discouraging idea of the value of scientific methods."

Huntington further argued that "the method of equal proportions . . . has been mathematically shown to have no bias in favor of either the larger or the smaller states." This claim is very strange, if not downright dishonest. His paper in the *Transactions of the American Mathematical Society*, published only a few months previously, had shown unequivocally that the H-H method is equivalent to rounding at the geometric mean. And rounding at the geometric mean clearly favors small states. Did the professor feign ignorance or did he really not know any better?

One way of proceeding would have been to refer the question to a knowledgeable institution that would check out the matter and prepare an independent expert report. Indeed, Willcox had suggested exactly this in hearings before Congress. Unfortunately, the American Political Science Association, which would have been well placed to arbitrate between the competing methods, refused to get involved. In a glaring display of ivory tower mentality the organization's secretary wrote that his association "has the feeling that it ought not to undertake to decide a question of this sort." Apparently the association felt that studying and criticizing how public figures make tough decisions is all right, but taking a stand is not.

In February 1929, Willcox weighed in at *Science* for the first time. His manner was more gentlemanly than was Huntington's. Rather than attacking his opponent personally he dealt with the problem at hand. The real danger loomed that the apportionment, first enacted in 1911 and unconstitutionally extended in 1921, would stay in effect for another ten years, until 1940, unless Congress decided to change it. Willcox wanted to deal with the big question instead of wasting time arguing about fractional seats. Hence, the objective was not to find the best or the fairest method but one that was workable and that would be acceptable to Congress. He expressed his goal succinctly: "What method is likely to give the bill the best chance of passing Congress?"

Convinced that his method of major fractions stood the better chance on this score, he could not but engage in a bit of demagoguery himself. Would the congressmen accept a "new, untried method" that is difficult to explain to laypeople, he wondered? Or would they prefer a method that had already been used in 1911 and involved no more than rounding fractions down if they are below 0.5, and up if they are above 0.5? Reminding the readers of a failed attempt to enact a previous apportionment law and eager to avoid a similar flop this time, he affirmed with selflessness that seems only slightly affected: "I would gladly abandon my preference . . . if I thought another method had a better chance of acceptance by Congress and the country." As for publishing in a recognized scientific journal, as Huntington demanded, Willcox did not care one bit. His work was not meant for the delectation of academics but as a service to Congress. "The judgment of the average representative or congressional committee is of far more importance than that of any group of scholars."

The riposte came after barely four weeks. Squeezed between "Geological Work in Tonga and Fiji" and "Importation of Cinematographic Films," a short note by Huntington appeared in *Science*, in which he outlines alleged errors in Willcox's paper. Huntington must have read the piece with a magnifying glass. Gleefully he jumped on a slightly ambiguous sentence in the paper. Willcox had written that the states' populations, divided by an appropriate divisor, give a series of quotients which, when rounded up or down, give each state's number of seats. "The whole series would add to 435." Deliberately misunderstanding "the whole series" to mean "the series of quotients," while Willcox had obviously meant "the series of representatives," Huntington termed the arguments in the paper crass misstatements of the mathematical facts, and the description of Willcox's method as grotesque. The note ends with a punch below the belt. "It appears to be only by evasive arguments like these that the method of major fractions can be defended." Ouch!

Willcox did not take the affront lightly. Handing copies of the ambiguous passage to a class of thirty undergraduates he asked them their opinion, without letting on the reason for the inquiry. Three-quarters of the class thought "the series" meant "series of representatives," only one-quarter interpreted the phrase to mean "series of quotients." Willcox reported the results of his inquiry in *Science* in March 1929 as if it had been a scientific experiment of great importance. "It is hard to understand how a scholar of the position of Professor Huntington could have given my words the meaning he did," he comments and closes his paper with the remark "Hitherto I have not answered Professor Huntington's personal attacks but this case is so clear and typical that I have made an exception."

The debate raged throughout 1928 and the first half of 1929. If nothing else, it served to show that politics in science could be every bit as bitter as politics in Congress. Huntington stepped in again in May 1929, but by that time the question had become moot.

It had become obvious by early 1929 that feuding politicians would be unable to come to an agreement and that academics, left to their own devices, did not behave any better. So Congress turned to the one institution that could save it from the impending imbroglio. It comprised experts, pundits in their fields, who could be counted on to settle any scientific question in an unbiased manner, uninfluenced by partisan or nationalistic considerations: the National Academy of Sciences.

So when Congressman Ernest Gibson of Vermont remarked that "the apportionment of representatives . . . is a mathematical problem," and posed the question "why not use a method that will stand the test . . . under a correct mathematical formula?" the Speaker of the House, Nicholas Longworth of Ohio, decided to put an end to the discussions by requesting the National Academy of Sciences (NAS) to decide on the appropriate method of apportionment.

Founded in 1863 by Abraham Lincoln, the academy's task is to give advice to the federal government and to the public about the impact of scientific and technological issues on policy decisions. As mandated in its Act of Incorporation, the NAS must "investigate, examine, experiment, and report upon any subject of science or art," whenever any department of the government asks it to do so. The members of the NAS—professors at universities, scholars at research laboratories, scientists in private companies—work outside the framework of government to ensure their independence. So essential has the academy's service to government become that over the years Congress and the White House have repeatedly issued legislation and executive orders that reaffirm its unique role.

In contrast to the American Political Science Association the NAS did take up the gauntlet. A commission was created that was to decide on the best method to use for apportionment. The blue-ribbon study group was comprised of the three mathematicians Gilbert A. Bliss from the University of Chicago, Ernest W. Brown from Yale University, Luther P. Eisenhart from Princeton University, and the group's chairman, the biologist and geneticist Raymond Pearl from Johns Hopkins University in Baltimore. Among the mathematicians, Pearl was the odd man out. However, he was well versed in statistics, having spent a year in England with Karl Pearson, the founder of the world's first university statistics department at University College in London. It was a comfort to the members of the House, indeed to the citizens of the United States, to know that such high-powered scientists from Chicago, Yale, Princeton, and Johns Hopkins would be dealing with this problem even though it involved nothing more than basic arithmetic.

The report started out innocently enough. If apportionment were computed by simply dividing each state's population by the total population, it stated, the number of representatives would in nearly all cases consist of a whole number and a fraction "as, for example 7.3." (The latter remark was apparently added so that the report would be understandable even by the most obtuse reader.) Now if fractional voting were permitted in the House there would be no problem, the committee members remarked; each state would receive the exact number of representatives corresponding to the whole votes, plus—this was an innovative idea—an additional representative with the fractional vote. The latter would be an incomplete representative, so to say. But the Constitution did not provide for fractional voting in Congress. Hence this was a nonstarter.

Sharp as razors, they deduced that it was necessary to reach a solution to the apportionment problem in whole numbers. But by setting this condition, the mathematical nature of the problem was altered fundamentally. "It should be understood that frequently a problem in applied mathematics may have no unique solution, for the reason that the data initially given do not completely characterize the solution mathematically. In such cases a solution must be chosen for other than mathematical reasons among those which are mathematically possible."

The committee members went hunting for reasons other than mathematical ones. They considered all apportionment methods then known that avoided the Alabama Paradox and did lead to workable solutions. There were five: those proposed by Webster, Adams, Jefferson, and Hill, as well as another method suggested by James Dean, a mathematics and physics professor at the University of Vermont during the 1820s. I have not spoken about Dean's method because it was never used, except in a legal challenge by the state of Montana in 1991. (It consists of adjusting the value of "population per seat" for the United States as a whole in such a manner that "population per seat" for each state is as close as possible to that value.)

The four men liked the idea of minimizing relative differences; thus from the outset the odds were stacked in favor of H-H, which does exactly that. Sure enough, the report culminated in the assessment that the method of equal proportions is preferable to the others because it minimizes relative differences. "After full consideration of these various methods your committee is of the opinion that, on mathematical grounds, the method of equal proportions is the method to be preferred."

The way the professors arrived at their conclusion is somewhat reminiscent of circular reasoning. First they stated their preference for minimizing relative differences, next they chose the method that was designed to do just that. Not much left to decide then, really. They did concede that if the *absolute* difference in the number of constituents required for a representative should be minimized, a different method would be optimal. And if the absolute difference of the inverse of this number, that is, representatives per constituents, were considered, yet a different method would be optimal. "Each of the other four methods listed is, however, consistent with itself and unambiguous," they concluded.

But the professors gave a further justification for their choice. They preferred H-H to the other methods because "it occupies mathematically a neutral position with respect to emphasis on larger and smaller states." Signed: G. A. Bliss, E. W. Brown, L. P. Eisenhart, Raymond Pearl, Chairman.

By "neutral" the four committee members did not mean that H-H is unbiased—we know that the geometric mean penalizes larger states—but rather that in terms of bias H-H occupied the middle place among the five methods. Modern scholars remarked that it was fortunate that the committee had an odd number of methods to start with; otherwise it would have been difficult to identify the best method as the one in the middle.

Huntington could breathe a sigh of relief; his method of equal proportions had been vindicated by the committee. In a recapitulation of the NAS's findings he gives a gleeful account of the committee's report in *Science*. "All controversy surrounding the mathematical aspects of the problem of reapportionment of Congress should be regarded as closed by the recent authoritative report of the National Academy of Sciences," he states at the outset, and then he proceeds to take another swipe at Willcox's "complicated and artificial" method. "The hold which this now obsolete method still maintains on the imagination of many congressmen is due mainly, it appears, to a misconception."

In another punch below the belt, Huntington did not forego the occasion to repeat the allegation that Willcox had meant to add a series of quotients instead of congressmen. But all this may now be forgotten because "the National Academy of Sciences confirms that . . . the method of equal proportions . . . is logically superior to the method of major fractions." The method of equal proportions received the unanimous stamp of approval by the NAS because it allocates seats in Congress in a way that minimizes disparity. An apportionment made according to it cannot be improved, in the sense that the shift of a seat from one state to another will increase disparity. Huntington writes, Willcox's "useless complications . . . are completely done away with in the modern theory which provides a simple and direct test for the settlement of any dispute between two states." The diatribe ends with a blow at the regrettable failures of recent history. "The purely political attempts which have been made to retain the obsolete method of major fractions in current legislation have proved to be a serious menace to the whole reapportionment movement."

Did we say that the NAS committee had issued its report unanimously? We did, and formally this is not incorrect. But a few years after the report was published, Willcox dropped a dark hint. He suggested that the report had not been issued unanimously. How could that be? The four members of the committee, Bliss, Brown, Eisenhart, and Pearl had all signed the report. Did Willcox mean to insinuate that one of the professors had been coerced? This was preposterous and any such suggestion should immediately be dismissed. Huntington was appalled by the suggestion. "Any attempt to show . . . that these signatures do not mean what they say, is a gross insult to these distinguished scholars," he wrote with deep indignation. So was the non-unanimity a figment of Willcox's imagination?

It was not, and the files at the National Academy of Sciences contain a dark secret. Nobody was coerced but what is missing from the historical record is that the committee had originally included a fifth member, the Harvard mathematician William Fogg Osgood. What had happened?

Osgood had been recruited to the committee with Bliss, Brown, Eisenhart, and Pearl and had joined in the drafting of the report's first version. It was already clear that the committee would endorse the method of Huntington, his colleague at Harvard, but as work progressed, Osgood became disillusioned. He felt that the Harvard method was not given strong enough support. Finally, he had a change of heart about his collaboration. He announced his resignation in a telegram of January 30, 1929 to the NAS home secretary. It read:

> Report Committee has been so far weakened through introduction of irrelevant material that I find myself unable to subscribe to the revised form STOP in order not to obstruct the proceedings I beg to be relieved from further service on the committee.

Two days later, in a telegram to committee chairman Pearl, he provided more detail:

> My resignation from the Committee was intended to expedite a unanimous report from the other four members STOP As you have sent me third draft, I am glad to comment as follows STOP I should to add [sic] to paragraph five QUOTE but the present problem does not belong to this class UNQUOTE or omit the whole paragraph STOP in nine I should wish to omit the last sentence STOP You must realize that I have consistently adhered to first part of report as drafted and unanimously accepted in Baltimore and circulated STOP My objections are to new material introduced since then at the instigation of President of the Academy. This new material in my opinion robs the report of its essential meaning STOP My resignation therefore stands.

Three days after that telegram, Raymond Pearl gave an account of the situation in a letter to the president of the National Academy of Sciences:

This report has been signed by four members of the Committee, namely Bliss, Brown, Eisenhart, and Pearl. The fifth member of the Committee, Prof. William F. Osgood, resigned from the Committee during the period of its work, for reasons which he has stated to you. I am informed by him that he wishes his resignation to stand, and desires to take no part in the further work of the Committee from the time his resignation was dated. I see no alternative, therefore, except for you to accept his resignation and consider the report to have been drawn by the four members of the Committee who signed it. I am very sorry that Professor Osgood felt obliged to resign. I did everything in my power to induce him not to do so, as did every other member of the Committee. Furthermore, every member of the Committee went just as far as he consienciously [*sic*] could to meet Professor Osgood's views at every point.

Willcox was not so naive as to believe that Osgood's resignation came about because the Harvard professor agreed with the Cornell method. To the contrary, it was obvious to him that Osgood had resigned because the committee's support for the Harvard method was not strong enough. He had accused the president of the NAS of having watered down the committee's conclusions to such a point that he could no longer agree with them. So actually Osgood's continued collaboration with the committee would not have done anything to further Willcox's cause.

For a long time, nothing further was heard on the subject from Willcox. Only in 1941 did he reveal that he had received a letter from Raymond Pearl, the NAS committee's chairman. In it Pearl belatedly vindicated Willcox's method. "Your efforts I heartily and unreservedly endorse. . . . In my opinion you are now doing a real service in bringing [the method of smallest divisors] to the fore."

If—in the wake of the NAS report—Congress would at long last decide on the method to be used, then finally, after twenty years, a new apportionment could be effected. Not everybody was happy with that prospect, however. Whichever apportionment method would be used, it would certainly be to the detriment of the rural states whose populations had diminished significantly. They were determined to fight tooth and nail against the looming danger to their interests. If they once again could torpedo attempts to reapportion the House, there was a chance that the present apportionment would be kept at least for the next ten years. Senators and congressmen from the rural states railed and fulminated against the "unholy, unrighteous, and unjust" method that "would run us down under the wheels, crush us by the system of major fractions . . . ruthlessly and without mercy." Their main aim was not just to derail W-W, but also to prevent any kind of agreement.

Notwithstanding their resistance, in the summer of 1929 Congress passed a bill that stipulated that the president would send the census data to Congress, together with two apportionment suggestions, one computed according to the W-W method, and another computed according to the H-H method. If Congress could not decide between the two methods the W-W method used in 1911 would automatically be employed again.

But then a remarkable if not wholly surprising coincidence occurred. After the census data were collected and counted, and the apportionments were computed and compared, it turned out that the W-W and the H-H methods gave identical results. Congress did not have to decide which method to use, nobody was crushed under the wheels, and apportionment could be effected in 1931 in idyllic harmony. Everyone breathed a sigh of relief and Congress could rest for the next ten years—at least concerning the vexatious apportionment issue.

The next time around, Congress was not so lucky. Following the 1940 census, everything came tumbling down again. When President Franklin D. Roosevelt presented the apportionments according to the two methods, forty-six states did indeed get the identical number of seats. But Arkansas and Michigan did not. Under the H-H method Arkansas would receive seven seats and Michigan seventeen, under the W-W method Arkansas would receive six and Michigan eighteen seats.

That was enough to get the controversy going again. And not only the two states joined the fight. Since Arkansas was a solidly Democratic state and Michigan usually voted Republican, the question became a partisan issue for the House. All of a sudden, the correct answer to a mathematical question boiled down to whether you were a Democrat or a Republican. The former preferred H-H, the latter W-W. (For what follows it is important to know that in 1941 Democrats held the majority in the House.)

TABLE 10.2A AND B
Comparison between H-H and W-W (1940 census)

(A) Huntington-Hill method (equal proportions)

Total population: 7,205,493. 24 seats to be apportioned. Population per seat = 300,229

				Option 1			Option 2		
State	Population	"Raw" seats	Allocated seats	Citizens/seat	Diff	Allocated seats	Citizens/seat	Diff	
AR	1,949,387	6.493	6	324,898	11.26%	7	278,484	11.02%	
MI	5,256,106	17.507	18	292,006		17	309,183		
Tot.	7,205,493	24.000	24			24			

Since the relative difference in Option 1 (11.26%) is higher than in Option 2 (11.02%), the latter is preferred.

(B) Webster-Willcox method (major fractions)

Find a divisor such that the rounded seats add up to 24: 300,000. (Actually, any divisor between 299,906 and 300,349 would work.)

	Population	"Raw" seats	Rounded seats
Arkansas	1,949,387	6.498	6 (rounded down)
Michigan	5,256,106	17.520	18 (rounded up)
Total	7,205,493	24.018	24

CHAPTER TEN

On February 17, 1941, the House started debating the issue. As stipulated by a resolution that the House had passed a year previously, the clerk of the House had presented the proposed apportionments among the states, based on both the Cornell and the Harvard method, on January 8. Unless the House took some action on the matter within sixty days, that is, by March 9, the apportionment would automatically be effected according to the old W-W (Cornell) system of major fractions. Thus it was high time for the Democrats to become active; otherwise, Arkansas could wave its seventh seat good-bye at least for the next ten years.

The discussion was kicked off by J. Bayard Clark from North Carolina. He pointed out that Michigan would gain a seat to the detriment of Arkansas if the method of major fractions were employed—even though the population of Arkansas had increased faster than that of Michigan. "The House would not want to let some mathematical formula result in an inequity or an injustice of this sort," he exclaimed, capitalizing on the congressmen's subliminal distrust of mathematics. (Little did it matter that his statement was quite untrue anyway: between 1930 and 1940, the population of Arkansas increased by 5.1 percent while the population of Michigan grew by 8.5 percent.) Joseph W. Martin from Massachusetts interjected: "We are trying to upset what we agreed upon last year. . . . Why should we change the understanding just because of a particular advantage to any one State?" Clark would have none of that. "The House ought not to permit any mere mathematical formula to defeat equity," he thundered from the floor, again driving home his point about the alleged vagaries of mathematics.

Out of breath by now, Clark yielded ten minutes of his speaking time to Congressman Ed Gossett from Texas. Gossett put his colleagues at ease. "We should not go into the intricate mathematical and geometrical formulas necessary to understand these various methods," he suggested, alleviating their innate fear of mathematics. Instead, he proposed to take the National Academy of Sciences's word for what the best method is. "Scientific opinion has concluded . . . that equal proportions tend to equalize the size of congressional districts." The Huntington-Hill method thus vindicated, Arkansas would keep its seventh seat.

Michigan would not take this lying down. Making the case for his state, Congressman Earl C. Michener started out in a conciliatory tone. "None of the methods is perfect," he remarked soothingly, "and it is the rule rather than the exception that experts do not agree." Conceding that "everybody is sincere and thoroughly convinced that his philosophy should be adopted by the Congress in the name of justice and equity," he went on to admit that there is no absolute answer. In this case, rather than making unwarranted changes, it was preferable to keep the method that had been decided upon previously. "No one knew at that time where the shoe was going to pinch." And then he went for the jugular. "The professors, statisticians, and mathematicians presented their arguments and after most careful consideration by the Census Committee and the Congress itself, the method of major fractions was adopted as preferable under all circumstances," he exclaimed, conveniently ignoring the National Academy of Sciences' contrary conclusions.

Then it was Arkansas' turn again. "There is nothing sacrosanct about the various methods used," the state's representative David Terry conceded. The House must simply seek a method of apportionment that would "settle this vexing decennial problem in a way that would be most equitable and fair to all the States of the Union, large, medium, and small." Then he got to the point. No longer content with hiding behind niceties, Terry cut to the chase. "The Republican side of the House is making a desperate effort to control this body," he exclaimed to the applause of his fellow Democrats. Ezekiel Gathings from Arkansas feigned surprise at the surge in activity of the lawmakers from Michigan. Why this protest, all of a sudden? After all, when the Committee on the Census debated the issue, no representative from Michigan had ever bothered to attend in order to argue his state's case. This was too much for Jesse P. Wolcott. He and his colleagues from Michigan had been under the impression that everything had already been settled, that is why they had not shown up. But Gathings did not let this go by. Everybody had known for weeks that the matter was being discussed. Apparently the gentlemen from Michigan did not really want the seventeen congressional districts in their state disturbed, he concluded. "Oh just a minute," an enraged Wolcott shot back. "What does the gentleman think we are fighting for? We are not sitting here with our mouths open just for the fun of it. . . . Michigan is entitled to the seat and we are putting up a fight for it."

The debate became even more incensed. Fred C. Gilchrist from Iowa introduced a totally new aspect into the discussion. He did not care which method would be used, because Iowa was going to lose a seat whatever method was going to be used. He had a different cat to skin: non-natural-ized immigrants, at that time referred to as aliens. Using a chart prepared by the Immigration and Naturalization Service, he pointed out that of the nearly 5 million noncitizens living in the United States, one-quarter lived in New York State and another 11 percent in California. Thus four extra congressmen were allocated to New York and two to California, on the basis of aliens living in these states. "No reason can be given why any man who is born in a different country and who does not think enough of America to become a citizen . . . should be counted in determining the ap-portionment of representation." Worse still, Gilchrist reminded his listen-ers, "some of them hid behind foreign flags and refused to fight for Amer-ica, but stayed at home and received $10 or $12 or $15 a day as wages while our own boys gave up their jobs and got from $1 to $1.10 per day, and many thousands of them never came home at all, but were killed in the shambles of European battlefields." Then he turned to his real gripe. "A very large majority of these aliens are located in the cities, and this fact tends to take away from the rural sections and farm sections a fair and equitable share in the control of legislation." Thus, the rural states were doubly punished by the aliens, Gilchrist lamented. Not only did they take the jobs of American boys, but also by settling in towns and cities they prevented the rural states from being correctly represented in Congress.

As the discussion progressed, others joined the fray. Leland M. Ford from California had his say and so did August H. Andresen from Minne-sota; John R. Kinzer from Pennsylvania made his point and A. Leonard Allen from Louisiana made his. Carl T. Curtis from Nebraska gave his opinion and so did many others. And so it went, on and on. Of course, the whole exercise was ultimately futile. No argument could justify one method over the other since the question boiled down to whether one wanted Ar-kansas or Michigan to get the additional seat. Appeals to fairness and eq-uity carried no weight. Neither did the invocation of mathematical author-ities or institutions, be it the NAS, the Brookings Institution, or the Census Bureau, since for every argument a counterargument could be found that would justify—always for the most lofty reasons—whichever method one wished.

While more or less polite banter took place in the House, tempers were rising among the academics again. This time, much of their learned de-bate was vented on the pages of *Sociometry*, a journal founded in 1937 and devoted to matters of social psychology or, if you will, psychological sociology. (*Sociometry* appeared until 1977 when it changed its name to *Social Psychology*. Today it is called *Social Psychology Quarterly*.) Huntington started the debate with an article titled "The Role of Mathematics in Congressional Apportionment." Prefaced by an editor's note that "Huntington's suggested method of apportionment . . . is before Congress as we go to press," the paper purports to give a mathematical justification to the Harvard method. A mathematical theorem is either true or false, Huntington asserts, and then produces a "theorem of equal proportions." Actually this seems to be the first time that anybody ever spoke of such a theorem, but Huntington claims that "the truth of the theorem is vouched for by the unanimous Report of the Census Advisory Committee . . . and the unanimous report of a committee of the National Academy of Sciences." Such weighty support certainly should sway Congress, but further inspection reveals that Huntington was being disingenuous by appealing to the readers' respect for mathematical theorems and their proofs. (Note how Huntington tried to curry favor with the readers of *Sociometry* by appealing to their respect for mathematics, while Congressman J. Bayard Clark had tried to convince his colleagues by appealing to their disrespect for mathematics.)

What Huntington did was to posit his favorite question—does the reallocation of a seat from one state to another minimize the percentage inequality of the congressional districts in them?—as an appropriate test of a good apportionment. Then he stated, as a theorem, that the Harvard method satisfies this test. This is quite sneaky; since everything depends on the assumption, the argument is circular. If one accepts the test, the Harvard method is acceptable. If one does not, everything is open. Hence, the "theorem" says nothing about which method is the better one.

To hide the obvious shortcoming in his argument, Huntington takes refuge in demagoguery. He deplores that some "professional politicians . . . largely influenced by Professor Willcox of Cornell, sharply resent the intrusion of mathematical theories in a field which they regard as purely political, not at all mathematical. Professor Willcox flatly rejects (or, more accurately, deliberately ignores) the mathematical theorems cited above." He further accuses Willcox, who had maintained that the Harvard method was difficult to understand while the theory underlying the Cornell method is persuasive to the nonmathematical mind, of insulting Congress. "This is perhaps the first time in history that advocates of any measure have openly accused the Congress of the United States of being unable to multiply and divide," he proclaims.

Willcox would not let that pass. "So long as [Huntington's articles] were safely immersed in the catacombs of Public Documents I ignored them. But now that [they] are accessible to uninformed persons whose opinions I value . . . the antediluvian octogenarian must turn away from more congenial tasks to answer him," the by then eighty-year-old professor wrote in a reply in the same journal.

Choosing the proper method of apportionment is a political one, Willcox maintained, since the choice is made by a political body for political motives. "The force behind the bill is not a tardy conversion of Congress to the method of equal proportions but a discovery by the leaders of the majority that [a switch of method] will transfer a seat from Michigan to Arkansas." Hence the problem at hand was not "a choice between two methods. In reality it turns out to be a choice between two parties." Allowing a conversion to the Harvard method now, Willcox fulminated, would open a Pandora's box of future trouble, with the party in power switching apportionment method to suit its needs.

The basic problem is "to reach a result that is as near as may be to an exact apportionment," Willcox remarks and then correctly points out that this begs the question "how is 'nearness' to an exact apportionment to be measured?" As the NAS report had pointed out twelve years earlier, much depends on whether "nearness" is defined in absolute terms or as a percentage. The proponent of the Cornell school then stoops to some demagoguery of his own, albeit in a subtler and more poetic fashion than the advocate for the Harvard school. Huntington's articles "reveal the writer as a modern Don Quixote roaming in an unreal world where he tilts against a Congressional windmill the structure of which he fails to understand and the forces governing which he has been unable to influence." Concluding his reply, Willcox regrets that Huntington did not let sleeping dogs lie. "In reviving the struggle of 1929 he has now innocently but nonetheless efficiently dragged the problem of apportionment back into the quagmire of politics from which his academic opponents had long struggled to extricate it and has left it for his successors and mine in a shape far more complicated and menacing than it would have been had he never touched it."

To make a long story short, Willcox was not convincing enough and President Roosevelt did not abandon the Democratic Party. On November 15, 1941, without much consideration as to the merit of the method, he signed into law "An Act to Provide for Apportioning Representatives in Congress among the Several States by the Equal Proportions Method," thus decreeing that Huntington-Hill would henceforth be used to apportion Congress. The Harvard method had carried the day and the gentlemen from Arkansas could breathe a sigh of relief. They got their seventh seat. And the Democratic majority in the House increased its majority by one.

But misgivings about the crude manner in which H-H had been adopted abounded in the House, and in 1948 the question was put to a scientific test once again. The confused Congress turned to the National Academy of Sciences for help, asking it, once again, to investigate which apportionment method the Congress should use. The NAS formed another committee. This time around, the committee was purely a Princeton affair and the committee was, if anything, even more blue ribbon than the first one. John von Neumann from the Institute of Advanced Study (IAS) was its chairman. An émigré from Hungary, he is today considered one of the most important mathematicians of the twentieth century. (The IAS provided him—as well as his colleagues Albert Einstein, Kurt Gödel, and other refugees from Nazi-Germany—a quiet place to further human knowledge, undisturbed by banal duties such as teaching students or supervising doctoral candidates.) Another member of the committee was Marston Morse, von Neumann's colleague at the Institute for Advanced Study at Princeton. The committee's third member was Luther Eisenhart from Princeton University who had already been a member of the NAS's first committee two decades earlier. He was chosen to be the chairman.

The three mathematicians went to work. They had been asked by the president of the NAS—who, in turn, had been asked by the Speaker of the House—to report on any new developments in the mathematical aspects of the apportionment problem since the previous report of 1929. Actually, there was not much new to report. The only paper written since then, and worthy of consideration by the committee, was one prepared by Walter E. Willcox for a meeting of the International Statistical Institute. In it he proposed a new apportionment technique: the modern House method. But even that was no innovation. Upon inspection, it turned out that the modern House method was no more than a method that had already been considered in the NAS's previous report, albeit under the name method of smallest divisors. So the committee could limit itself to covering substantially the same ground as its predecessor.

After again lamenting the fact that "fractional representation has not been so far introduced," Morse, von Neumann, and Eisenhart subjected the various methods of apportionment once more to intense scrutiny. They compared the H-H method of equal proportions that had already been considered superior in the previous NAS report with the competing methods. This they did by subjecting the methods to the test whether the differences between the numbers of citizens required for a seat decrease if any of the other methods are used. One member of the committee—the report remains silent on his identity—was designated to work out the comparisons algebraically. He did, and it comes as no surprise that the mathematical paperwork confirmed the conclusion of the report that had been submitted to Congress nineteen years previously. "In the above four comparisons EP [the H-H method of equal proportions] scored decisively in each case," the report concluded. This was not altogether mind blowing since the new committee was not about to disavow its forerunner, especially since one-third of the members—the committee's chairman Eisenhart—provided continuity.

But there remained a loose end. After all, Willcox had put forth the modern House method a.k.a. the method of smallest divisors and "this report would not be a reply to the request of the National Academy if it did not analyze the recent paper of Professor Willcox." Indeed, apart from repackaging and changing the name, Willcox believed he had found a further justification for his modern House method. He suggested comparing the different apportionment methods not just between pairs of states but also between *all* states. What is the difference, he asked, between the largest population and the smallest population required for a seat? Willcox called this new variable, computed over all forty-eight states, the "range." The apportionment method that results in the smallest range would be the preferred one.

Inspecting the ranges for the 1940 census according to the different apportionment schemes, Willcox found that it was smallest under the modern House method. It should therefore be the preferred method, he argued. But his analysis had a snag. True to his suggestion that the comparisons encompass all states, Willcox had included Nevada in his calculations. This state, with a population of only 110,247, was too small to receive representation in the House by force of its size; it had come by its seat only because of the law that each state was to have at least one representative. Hence, the number of citizens required for Nevada's single seat (110,247) was exceptionally low, especially when compared to South Carolina, which received six seats for a population of 1,899,804 or 316,634 citizens for a seat.

Morse, von Neumann, and Eisenhart argued that Nevada should be excluded from any comparison since the allocation of this state's seat was not based on any of the apportionment methods. And—here comes the knockout punch—once Nevada is excluded from the computations, the range is slightly lower under the H-H method. It would have been the other way around after the 1930 census. Thus, including or excluding this or that state could completely change the result. Exasperated, the committee members threw their hands collectively up into the air. "To use the language of gun-fire," they wrote in their report, "the test of minimum range makes the evaluation of a method dependent upon a few eccentric shots, and in this sense is a random determination of value." Unwilling to succumb to such randomness, the committee stuck to its preference of the Huntington-Hill method.

Such was the state of affairs toward the middle of the twentieth century. Whatever its advantages or disadvantages, Huntington-Hill a.k.a the Harvard method has been the method of choice for Congress ever since. Of all methods considered, it represented the middle way and therefore was the most acceptable. Nevertheless, the whole affair left a bad taste because on the theoretical level the question had not been settled at all. One could not get rid of the arbitrariness to which the mathematical imprecision of the rounding process gives rise. It was only partly tongue-in-cheek that some wags suggested explicitly introducing randomness into the apportionment method. They proposed a roulette-based method to distribute fractional seats. The width of each cell on the roulette wheel would correspond to the size of each state's fractional seat and—*les jeux sont faits*—the state into whose cell the marble falls, gets the leftover seat. This is less absurd than it sounds because, on average, the method is absolutely unbiased; in the long run, every state gets its fair share of left-over seats. But since apportionments take place only once every ten years, it would take a very long run indeed for the odds to average out. And as the eminent English economist John Maynard Keynes once remarked, "in the long run, we are all dead."

BIOGRAPHICAL APPENDIX

Walter F. Willcox

Many social scientists consider Willcox the "father of American demography." He started the teaching of statistics at Cornell. Under the heading "applied ethics" he offered "an elementary course in statistical methods with special treatment of vital and moral statistics" in the department of philosophy in 1892. It was one of the earliest courses in social statistics in the United States. Willcox's main achievement was the application of the still young science of statistics to the area of demography. By today's standards he was not a sophisticated statistician. According to Frank Notestein, a former student who would later become director of the Office of Population Research at Princeton University, he barely knew what mean, median, and mode were and used only simple methods instead of high-powered techniques. Then again, statistics was a new subject at the time, so this was not very surprising. Most importantly, he had a very healthy respect for data and everybody agreed that he was a great teacher.

Many scholars today accuse him of having been one of the proponents of scientific racism. Apparently Willcox believed in the racial inferiority of Afro-Americans—he still called them Negroes, in the same manner that he spoke of "girls" when we would today respectfully say "women"—and tried to explain the plight of black farmers with this alleged inferiority. Willcox was active well into his nineties and died at age 103. At various times, he served as president of the American Economic Association, president of the American Statistical Association, and president of the American Sociological Association. He was a prominent member of the highly selective and narrowly restricted International Statistical Institute, attending most of its meetings, be they in Tokyo, Warsaw, or Rio. He was even present at a meeting in Prague in 1938 that had to be cut short when Hitler invaded Czechoslovakia.

Joseph A. Hill

One does not expect the life of a statistician to be as adventurous as that of, say, an archaeologist like Indiana Jones, but Hill's vita seems to have been gray even by the modest standard that one would apply to a member of his profession. Hill studied at Harvard and then became a statistician for the government. "The nature of his work was such as to afford little opportunity for publicity," an obituary in the *Journal of the American Statistical Association* read and then went on to describe the "arduous and relatively thankless job of production and processing of statistics" that Hill performed throughout his life. "Relative anonymity was inherent in the nature of his job," the biographer continues, and then can do no more than to extol Hill's "necessary care and extravagant meticulousness." Wistfully, he adds that "this sort of service produces no renown." At a loss to say anything extraordinarily positive, he presents Hill's compilation of a "well-constructed statistical volume with clear and precise headings and titles" as a highlight of his "patient, imaginative, but unspectacular work."

His output, though wide-ranging—his statistical tractates cover crime, fecundity, insanity, migration patterns, child labor, marriage, and divorce—was quite unspectacular. "In his work relatively little of higher mathematical and statistical technique was required," the biographer recounts. Since one does not end an obituary without mentioning so much as a single outstanding aptitude, Hill is described as a master at "the art of extracting the last drop of legitimate meaning out of a body of statistics." But, commendably, he never let his quest for more information get the better of him. Hill squeezed informa-

tion from numbers while strictly adhering to an "unflinching recognition of what the figures do not and cannot prove." Especially praiseworthy in the eyes of the biographer was the self-discipline that kept Hill from "stretching the material to prove some pet theory." Not only that, but his publications were always provided with an "honest and precise statement of the margin of error." Truly a virtuous man.

Then there was, of course, the census. In fact, for a whole generation of budding statisticians Hill was identified with the work of the Census Bureau. He gave lectures on the census to the American Historical Association, the American Sociological Society, the American Statistical Association, the Federal Council of Churches, and he wrote about it for *Youth's Companion*, the *National Republic*, the *New York Times*, and the *Monthly Labor Review*. In addition, he authored numerous unpublished internal documents about this or that aspect of the census. As chairman of the Quota Board, he was instrumental in allotting immigration quotas to countries in the same proportion that the American people traced their origins to those countries.

Although far from the limelight, the behind-the-scenes work of this dedicated professional must not be belittled. His whole life was devoted to improving the quality of the material on which public policy is based. It was in this spirit of service to the public that he thought deeply about apportionment methods. His ideas on that subject were published in the *Congressional Digest*.

Edward V. Huntington

Born in 1874, Huntington was educated at Harvard, became an instructor at Williams College, and went to Europe to obtain his doctorate in mathematics in Strasbourg, then in Germany. Upon his return to the United States he began a career at Harvard, starting as an instructor and gradually making his way up the academic ladder to assistant professor, associate professor, and full professor. In contrast to most colleagues at math departments throughout the world, Huntington especially enjoyed teaching mathematics to engineering students, which earned him the additional title of professor of mechanics. World War I found him in Washington where he dealt with statistical problems for the military. Huntington's primary research interests were the foundations of mathematics, and he did important work on axiomatic systems in algebra, geometry, and number theory.

MEMBERS OF THE NAS COMMITTEES

Gilbert A. Bliss

Bliss, professor of mathematics at the University of Chicago, was born in 1876 into a wealthy family. His father was the president of the Chicago Edison Company, which supplied most of Chicago's electricity. But the family fell on hard times during the Depression and the young man had to earn his way through university as a professional mandolin player. After obtaining his PhD at the University of Chicago, Bliss went to the then stronghold of modern mathematics, the University of Göttingen in Germany, for a year. There he met the towering giants, Felix Klein and David Hilbert. World War I found him designing firing tables for the artillery. Bliss was best known for his work on the calculus of variations.

Ernest W. Brown

Brown was born in England to a farmer and lumber merchant. Educated at Cambridge, he moved to the United States when he was twenty-five years old. Brown first taught mathematics at Haverford College in Pennsylvania, and was appointed professor at Yale in 1907. His primary interest was astronomy, and he published important work on lunar theory and the movement of the moon. For example, he correctly ascribed hitherto unexplained wobbles in the moon's orbit to irregular changes in the Earth's period of rotation.

Luther P. Eisenhart

Son of a sometime-dentist, Eisenhart had been a precocious child. At Gettysburg College he excelled in his studies as well as in baseball. After obtaining his doctorate in mathematics from Johns Hopkins University, Eisenhart spent his entire academic career at Princeton University, beginning as an instructor at age twenty-four in 1900 and continuing through the ranks until his retirement, as Head of Mathematics, forty-five years later. His chosen fields of specialization were differential geometry and, after his retirement, Einstein's theory of general relativity. Apart from his research and teaching duties, Eisenhart was also very active in administrative matters. At different times he served as president of the American Mathematical Society, officer of the American Philosophical Society, president of the American Association of Colleges, and editor of the *Annals of Mathematics* and the *Transactions of the American Mathematical Society*. He was awarded seven honorary degrees for services rendered to mathematics and to higher education in general, and King Leopold III of Belgium made him an Officer of the Order of the Crown.

Raymond Pearl

Pearl, from Johns Hopkins University in Baltimore, was among the first scientists to apply statistical methods and procedures to biological problems. The author of "Breeding Better Men," he at first supported eugenics and was a dues-paying member of the American Eugenics Society. Later he turned against this pseudoscience and his criticism of eugenics, published in the influential magazine *The American Mercury*, made it into the national headlines. It also earned him the enmity of many biologists. But progressive views on a pseudoscientific aberration did not prevent him from holding racist and anti-Semitic views. He was proud about how Johns Hopkins University dealt with the problem of the Jews. "For a number of years past, very quietly and skillfully, means have been taken and are being planned for the future to keep down our Jewish percentage," he wrote to a friend at Harvard. The secret was not to use crude quotas, as Harvard tried, but to practice discrimination. After all, "whose world is this to be, ours, or the Jews?"

In 1927, Pearl was the center of a high-profile case of academic infighting. Offered the position of head of a research institute at Harvard University, Pearl immediately resigned his professorship at Johns Hopkins in order to accept the prestigious post. The step was somewhat premature because a faculty member at the same institute, who had felt slighted by Pearl's earlier attack on eugenics, appealed the job offer to Harvard's Board of Overseers. In a rare step, the offer was rescinded and Pearl had to ask Johns Hopkins meekly to reinstate him—which the university did.

William Fogg Osgood

Osgood, son of a medical doctor, was born in 1864, and as a young man first wanted to study the classics. But after two years at Harvard he was persuaded by his teachers to switch to mathematics. He finished his undergraduate studies brilliantly, coming in second among 286 students. After an additional year at Harvard to earn his master's degree, Osgood obtained a three-year fellowship that enabled him to study in Germany. He learned to speak German and spent his time first in Göttingen, where, like Bliss, he was taken under the wings of the towering figure of Felix Klein, and later at the University of Erlangen, where he received his doctorate. In Göttingen he married Theresa Anna Amalie Elise Ruprecht, the daughter of the owners of a local publishing house. The couple would have three children before their marriage broke up and ended in divorce.

At Harvard, Osgood became instructor, then assistant professor, and then full professor. During his three years abroad he had absorbed much European mathematics, and he was instrumental in bringing methods and techniques to the United States. Osgood had taken a liking to all things German, supported Germany during the First World War, and even took to emulating the mannerisms of a German professor. From 1905 to 1906, he served as president of the American Mathematical Society. Somewhat late in life, Osgood decided to tie the knot again. The already sixty-eight-year-old mathematician married a forty-year-old woman, Celeste Phelpes Morse, the divorced wife of his eminent Harvard colleague, the mathematician Marston Morse, about whom I will have more to say below. Upon learning of this liaison, Morse got a shock, and in the ensuing scandal Osgood retired from Harvard. For two years he taught in Beijing before moving back to Massachusetts. Today he is best remembered for his work on differential equations and the calculus of variations. He died in 1943.

John von Neumann

Jancsi, as he was called then, was born in 1903 and became a child prodigy. At his father's request, a well-to-do banker in Budapest who wanted a practical profession for his son, Jancsi enrolled at the Swiss Federal Institute of Technology in Zurich to study chemical engineering. Simultaneously, and secretly, he studied mathematics at the University of Budapest—in spite of a quota against Jews—albeit without attending any classes. He put in a presence only for exams, which he passed brilliantly. In 1926 he received a diploma in chemical engineering from the Federal Institute of Technology (ETH) in Zurich and a doctorate from the University of Budapest. There followed a study year with David Hilbert at the world-famous math department of the University in Göttingen. Von Neumann was already considered a genius, and everyone who met him recognized his superior intellect. "By his mid-twenties, von Neumann's fame had spread worldwide in the mathematical community. At academic conferences, he would find himself pointed out as a young genius," the biographer William Poundstone wrote. During the years 1930 to 1933 he held positions both in Germany and at Princeton University. After IAS was founded, he became one of its six original professors of mathematics. In 1937 von Neumann, now Johnnie, became an American citizen. He died of cancer at the age of fifty-four.

Von Neumann is considered the father of modern computers. Actually he fathered many disciplines and subdisciplines. His groundbreaking contributions to mathematics, quantum theory, economics, decision theory, computer science, neurology, and other fields are too vast to be listed here. We just mention two areas. As a consultant to the Manhattan Project he was instrumental in the development of the atom bomb in Los Alamos. He worked out the theory of "implosion" that proved to be the key to the success of Little Boy, dropped over Hiroshima, and Fat Man, dropped over Nagasaki. The idea was that explosives, shaped in a certain way, should surround a subcritical mass of plutonium. Just after detonation, the shock wave would turn inward, crushing the plutonium into a supercritical mass. It is for his association with the Manhattan Project (and for the fact that the cancer-stricken professor was confined to a wheelchair during the last months of his life) that Stanley Kubrick reportedly had von Neumann in mind when he created the character Dr. Strangelove in his 1963 film. Among the many honors von Neumann received are two presidential awards: the Medal for Merit in 1947 and the Medal for Freedom in 1956.

Marston Morse

Morse served as a soldier in World War I in France after earning his PhD in mathematics at Harvard and was awarded a Croix de Guerre for his outstanding service in the ambulance corps. After the war, he taught at Cornell, Brown, and Harvard, before joining the Institute for Advanced Study at Princeton. He is most famous for developing "Morse theory," an area in topology, that is, the study of shapes. Morse was awarded twenty honorary degrees and named a Chevalier in France's Légion d'Honneur. A member of the second NAS committee, Morse provided some continuity from the first, one might say, since he was the first husband of the second wife of William Osgood who had served on, and then resigned from, the first NAS committee.

MATHEMATICAL APPENDIX

Rounding at the Geometric Mean

In this chapter the question was asked, why rounding at the geometric mean should be related to minimizing relative differences in the number of people needed for a congressional seat. Here we present a proof of the astonishing statement that both procedures give identical results.

Let $p_1, p_2, \ldots p_n$ be the populations of the states, let d be the divisor, and let $a_1, a_2, \ldots a_n$ be the apportionment that results from rounding the ratios p_i/d at the geometric mean.

Then for every state i we have

$$\sqrt{(a_i[a_i - 1])} \leq p_i/d \leq \sqrt{(a_i[a_i + 1])}.$$

This is equivalent to

$$a_i(a_i - 1)/p_i^2 \leq 1/d^2 \leq a_i(a_i + 1)/p_i^2.$$

Since this holds for every i, it follows that for every i and j,

$$a_i(a_i - 1)/p_i^2 \leq a_j(a_j + 1)/p_j^2.$$

Now assume, by way of contradiction, that the apportionment $a_1, a_2, \ldots a_n$ does not minimize relative differences in the number of people needed for a congressional seat. In that case there exists a pair of states i, j such that state i is better represented than state j, and a transfer of one seat from i to j would lessen the relative inequality between them.

This means that

$$([a_j + 1]/p_j)/([a_i - 1]/p_i) < (a_i/p_i)/(a_j/p_j).$$

But this implies that

$$a_i(a_i - 1)/p_i^2 > a_j(a_j + 1)/p_j^2,$$

which contradicts the earlier inequality. Hence, the apportionment that results from rounding the ratios p_i/d at the geometric mean must minimize relative differences in the number of people needed for a congressional seat.

After: Peyton H. Young, *Equity in Theory and Practice*, Princeton University Press, 1995.

I would like to thank Daniel Barbiero, Manager of Archives and Records of the National Academy of Sciences, for making available to me the content of the documents quoted here. The documents are contained in the folder: NAS-NRC Archives, Central File: ADM: ORG: NAS: Committee on Mathematical Aspects of Reapportionment: 1928–29.

CHAPTER ELEVEN
THE PESSIMISTS

We now leave the matter of apportionment for a while and return to the troublesome problem of electing a leader. Remember Condorcet and his paradox? And how Lewis Carroll wrestled with it? Well the problem did not go away. Nor did it mellow with age. If anything it became more vexing. Enter Kenneth Arrow, Nobel Prize winner of economics in 1972 and one of the most important economists of the twentieth century.

An outstanding graduate student at Columbia University in the late 1940s, Arrow was thinking about his doctoral thesis. It was an exciting time for budding economists, observing and shaping subjects in the making. Arrow was caught up in these "heady days of emerging game theory and mathematical programming," as he would put it later. In the meantime he neglected his PhD thesis. He had high aspirations and his teachers and colleagues also expected a lot from him. But it was as if he was spellbound. No topic that he considered seemed sufficiently challenging. Even though his coursework had been completed at Columbia back in 1942, he was still short a thesis six years later. While everybody knew he was brilliant, the years passed without his putting pen to paper.

There was hope, however. Just a few years earlier, at the Institute for Advanced Study in Princeton, John von Neumann, together with Oskar Morgenstern, a refugee from the Nazis in Austria, had finished a thick primer that would become one of the most influential scientific works of the twentieth century. *Theory of Games and Economic Behavior*, published in 1944, was to have a profound influence on the further development of economics and political science. Based on only a handful of axioms, the theory contained in their book, henceforth called "game theory," ushered in the age of mathematical economics. What Euclid did for geometry, von Neumann and Morgenstern did for economic behavior.

One of the fundamental assumptions of their new theory was that each participant in a game has a so-called utility function. As we shall see presently, utility functions are a fundamental concept not only for the understanding of economic behavior but also of the Condorcet Paradox.

Attempts to explain economic behavior had started more than two centuries earlier, with work by the famous Swiss mathematician Daniel Bernoulli. In 1713 Daniel's cousin Nikolaus posed the following question: imagine a game in which you flip a coin and if it comes up heads, you get two dollars. If it comes up tails you flip again and if it now comes up heads you get four dollars. If it does not, you continue flipping until it comes up heads, the prize money doubling at every flip. How much would you be willing to pay in order to participate in this game? Most people would be willing to wager somewhere between two dollars and ten dollars.

But why so little? After all, the prize money could be enormous. If the coin comes up heads only after the tenth flip, the payoff would be 1,024 dollars, after the twentieth flip it would be more than a million and after thirty flips it would be a cool billion. Admittedly, the probability of getting nineteen or twenty-nine tails in a row, and heads only on the twentieth or thirtieth flip is very small. But the huge prize compensates for the small probability. In fact, Nikolaus Bernoulli found that the expected prize is infinite! (The expected prize is calculated by multiplying all the possible prizes by their probabilities, and adding the resulting numbers: $(1/2 \times 2) + (1/4 \times 4) + (1/8 \times 8) + \ldots = 1 + 1 + 1 + \ldots$, an unbounded sum.) So, once more, we have a paradox: if the expected prize is infinite, why is nobody willing to pay a thousand dollars to enter this game?

After thinking about the question for a while, Daniel came to a surprising conclusion: a dollar is not always worth a dollar. At first blush this statement may sound like a contradiction in terms, but upon further inspection it is not at all unreasonable. After all, a beggar who owns only a single dollar puts a high value on a second dollar, whereas a millionaire would hardly notice the receipt of an additional dollar. Hence the "utility" of money differs according to how much wealth one possesses. The utility of an additional dollar declines the more dollars one already has. Therefore, Daniel argues, one has to take into account not the expected prizes, but the expected utilities of the prizes.

Now the search was on for a suitable utility function. The requirements were that it grows—more is better than less, and even a rich person prefers more dollars to fewer dollars—but that the utility of an additional dollar decline with increasing wealth—the millionth dollar is valued less than the first. Hence, the two requirements on the shape of the utility function are that it always be increasing but to an ever lesser degree. One function that fulfills both requirements hand in glove is the logarithmic function: it always rises but at a decreasing rate. Hence Daniel posited it as a suitable utility function. Accordingly, the utility of a second dollar after the first is 0.3, but the expected utility of the millionth dollars is only 0.0000004. Subjecting the coin-flipping game to the computations, it turns out that the expected utility of the game's prize is four dollars. This amount, which accords with our intuition, is the amount that Daniel Bernoulli thought an average person should be willing to pay in order to participate in the game.

The exact form of the utility function is open to discussion, and different people have different utilities for wealth. Daniel used the logarithmic only as an example. But the principle has become clear. In 1738, twelve years after Nikolaus's death, Daniel's solution to the problem was published in the *Commentaries of the Imperial Academy of Sciences* of St. Petersburg. Henceforth the problem became known as the St. Petersburg Paradox. (To see how utility function and the requirements on their shape imply the existence of the insurance industry, see chapter 36 in my book *The Secret Life of Numbers*.)

*　　*　　*

Starting in 1948, Arrow spent the summers at the RAND Corporation in Santa Monica, the original nonprofit global policy research institute that would set the standard for all think tanks that followed. (Altogether more than half a dozen men who would eventually win economics Nobel Prizes worked at the RAND Corporation at various times. Apart from Arrow, there were Herbert Simon [1978], Harry Markowitz [1990], John Nash [1994], Thomas Schelling [2005], Edmund Phelps [2006], and Leonid Hurwicz [2007].) Game theory and operations research were hot topics at the corporation's headquarters. The Cold War was just gaining momentum and it was not surprising that the think tank was commissioned to study how game theory could be used to analyze international conflicts and strategy.

It turned out not be a straightforward assignment. If the Cold War were to be considered a game between the United States and the Soviet Union—albeit a serious one—what are the utility functions of the two players? Do collectives such as nations possess utility functions at all? Individuals do, but how can their preferences be aggregated into something to which game theory can be applied? All of a sudden, Arrow had found his thesis topic.

His epiphany at RAND had occurred in the summer of 1948; now he had to sit down and write everything up. In October, as soon as he got back to Chicago from his summer job, he started working seriously, continuing to write for the next nine months. By June 1949 a paper emerged and was published a year later under the title "A Difficulty in the Concept of Social Welfare" in the *Journal of Political Economy*. At first his advisors were perplexed. Nobody was quite sure what this was, and—whatever it was—whether it belonged to economics. But when Arrow finally presented his PhD thesis it was a bombshell.

Published as a booklet with the title "Social Choice and Individual Values" by the Cowles Foundation two years later, the thesis was hailed a "critical evaluation of democratic theory in general, as well as of economic policy and welfare economics in particular." Containing barely ninety printed pages, it ushered in the theory of social choice. Its importance can be ascertained already on page 1 of the preface; no less than five future Nobel Prize winners in economics are listed in the acknowledgments: Tjalling C. Koopmans (1975), Milton Friedman (1976), Herbert Simon (1978), Theodore W. Schultz (1979), and Franco Modigliano (1985).

Arrow points out that the simplest way for a collective to make decisions is either to have a single person or a small group of people dictate the choices for the community as a whole, or to let choices be imposed by a set of traditional rules, for example by a religious code. The former method is a dictatorship, the latter is a method guided solely by convention. Both are undesirable because they disenfranchise the individual. In contrast, a democracy knows of two methods by which individuals can participate in the collective decision-making process: they may vote to decide on political issues and they may use the market mechanism—work for wages, buy and sell goods and services—to make economic decisions.

But as we already know, in particular from the writings of the Marquis de Condorcet (chapter 6), the voting process may lead to a problem. Allowing the majority to decide may result in cycles. Let me remind the reader of the paradoxical situation by way of an example: Tom's ranking in the presidential elections in 2000 was Bush before Gore and Gore before Nader; Dick ranked the candidates Gore, Nader, Bush; and Harry preferred Nader to Bush and Bush to Gore.

Tom:	Bush > Gore > Nader
Dick:	Gore > Nader > Bush
Harry:	Nader > Bush > Gore

A majority (Tom and Harry) prefers Bush to Gore, another majority (Tom and Dick) prefers Gore to Nader, and another majority (Harry and Dick) prefers Nader to Bush. The society, composed here of Tom, Dick, and Harry, prefers Bush to Gore, Gore to Nader, Nader to Bush, Bush to Gore . . . lo and behold, we have a cycle!

One may find reason to argue with Tom, Dick, and Harry's political preferences, but as orderings go, each individual ranking is perfectly reasonable. Nevertheless, when aggregating the preferences into a communal ranking by majority vote, something quite unreasonable happens. Arrow concludes that the simple majority vote is not acceptable as a method of arriving at a social preference from individual preferences. He also showed by example that less simple methods that involve, say, putting weights on top-ranked choices and then summing the weights of the individuals' choices do not work either. So how can the preferences of individuals be amalgamated into a social preference schedule that would be capable of ranking the many alternatives facing society?

The short answer is, they can't. But let us do the long answer. At the outset of his thesis, on page 2, Arrow states the objective of his endeavor. He poses the question whether "it is formally possible to construct a procedure for passing from a set of known individual tastes to a pattern of social decision-making." In other words, if every individual has a utility function, how can these utility functions be amalgamated into a social utility function?

In order to be a reasonable method of aggregation, the procedure in question would have to satisfy certain natural conditions of rationality. The problem is that the utilities of two or more people cannot simply be added. They cannot even be compared. Let me illustrate with an example. When deciding what to have at the bar, Dwaine orders a glass of wine and Dwight orders a can of beer. Pressed about his preferences, Dwayne may say that according to his utility schedule, he attaches 5 "utiles" to wine and 3 "utiles" to beer. In Dwight's worldview, beer is worth 12 of some other units, say "yotiles," and wine is worth 8 "yotiles." Would it make sense to say that Dwight likes beer four times as much as Dwaine? Or to say two beers are worth 15 "somethings" to Dwaine and Dwight? No and no again. One cannot add, subtract, or even compare utiles and yotiles. The utilities that different people attach to drinks, or to any other goods, cannot be compared.

It is like measuring temperatures. Let us say that on a particular day the temperature is 26 degrees Celsius in Paris and 78 degrees Fahrenheit in San Francisco. Obviously, it would be quite wrong to conclude that it is three times as hot in San Francisco as it is in Paris. When measured to different scales, temperatures cannot be compared simply by comparing the numerical readings. In a similar manner, each person has a specific, individual utility scale that cannot be compared to any other person's utility scale.

Arrow was nothing if not formal and rigorous. To set things in motion he postulated two axioms about the way people make choices, and five attributes that reasonable social utility functions should possess. For such a broad and all-encompassing theory as the theory of social choice this is quite parsimonious. That is as it should be. One of the hallmarks of a good theory is that it requires few axioms. They are, after all, unproven assertions on which everything else depends. The fewer such prerequisites a theory needs, the more powerful it is since it explains more with less. And the fewer axioms there are, the lower the chances are that they contradict each other, which would make the theory inconsistent.

As an example of an axiomatic system, let me cite Euclidean geometry as a case in point. In the third century BCE, the Greek mathematician had postulated a set of five axioms from which all of plane geometry could be derived. He had also claimed that this is the smallest possible set; each of the five axioms is required. Nevertheless, a feeling persisted among mathematicians that one of them, the parallel axiom, which states that exactly one line parallel to another line passes through any specific point on the plane, may be superfluous. They believed that this statement—and therefore all of plane (not plain!) geometry—could be derived from the other four axioms alone. For centuries mathematicians attempted to reduce the set to just four axioms, only to realize in the early nineteenth century that the parallel axiom was, indeed, indispensable for plane geometry. Without it, something completely different appears on the scene: non-Euclidean geometry. For example, on a sphere like planet Earth "parallel" lines must cross somewhere. Take a point on the equator and draw a line through it, pointing exactly north. Now take a point 100 kilometers to the west of that line, also pointing exactly north. The two lines, seemingly parallel, will cross at the north and south poles. Hence, Euclid's parallel axiom is violated on a sphere. Thus, one obtains Euclidean geometry with the parallel axiom, other geometries without it.

If five axioms are required for all of Euclidean geometry and four axioms for all of non-Euclidean geometry, then postulating no more than two axioms to describe rational decision-making is certainly not excessive. Obviously, Arrow includes no unnecessary dead weight. So what are the two axioms, the two unproven assertions that are both indispensable and sufficient to derive all of social choice theory? The first axiom says that when presented with two alternatives a decision maker is always able to make a comparison between them. Either he prefers one alternative to the other, or the other to the one, or he is indifferent between the two of them. For once, apples and oranges *can* be compared, at least with respect to the utilities they provide any individual. Recall in this context the legend about Buridan's donkey that starved to death standing halfway between two identical haystacks. Obviously, the animal took Arrow's axiom to its extreme. By dying, rather than turning to one of the haystacks, the poor animal showed that it was quite indifferent between the two alternatives.

The second axiom concerns the matter that has been dogging us since we discussed the work of the Marquis de Condorcet: cyclical preferences. It is conceivable that someone prefers juice to milk, and milk to water, but then decides that he prefers water to juice after all. Well in a perfect world this must not be, and Arrow postulates that a rational person's preferences must be transitive. This is a fancy way of saying that preferences carry through; if juice is preferred to milk, and milk to water, then juice must be preferred to water. Thus, while the majorities of a group of people may exhibit cycles, in Arrow's scheme of things individuals are not allowed to do so.

The only requirements that Arrow imposes on decision makers are that they can always make a choice, and that these choices must not result in cycles. The two axioms are very plausible; it could not get much simpler than that. (However, see the appendix "The Axiom of Choice" for a subtle problem with the first of these axioms.) We may now move on to the social utility function. Arrow felt that the requirements he put on individuals' choices also make sense for group choice. How can one get from individual preferences to the collective choice of a group? Since utilities cannot be added, something more sophisticated is needed. A mechanism that aggregates individual preferences into a "social welfare function" must satisfy certain conditions in order to be acceptable. As we shall see, they are also quite reasonable. The conditions accord with common sense and with our intuition about fairness and the democratic process.

The first condition goes under the technical-sounding name "unrestricted domain." It says that there must not be any restrictions on the personal utility functions that are to be aggregated except for the two mentioned axioms. Hence, the mechanism should work for all possible combinations of individual preference schedules. As long as they satisfy the two axioms, no orderings should be excluded. As reasonable as this requirement sounds, it is not always satisfied. Some orderings may be excluded for cultural or religious reasons, or a constitution guarantees certain rights, even if everybody would prefer otherwise. Limiting the choices to two candidates, as in runoff elections, also violates the unrestricted domain condition. The more serious problem occurs, however, when the condition of unrestricted domain *is* fulfilled. If electors are allowed to rank alternatives in any way they choose—as long as the individual rankings are transitive–cycles may result. We saw this in numerous examples throughout this book. Charles Lutwidge Dodgson a.k.a. Lewis Carroll made the suggestion to break cycles by disregarding some voters' preference orderings. (See chapter 8.) This would be a violation of Arrow's first condition.

Arrow's next condition is called the monotonicity requirement. It says that if one individual raises the ranking of an option, while everybody else keeps it constant, society as a whole cannot react by reducing this option's rank. To illustrate, if society decided that lemonade is preferable to orange juice, and one individual now changes his preference from orange juice to lemonade, it cannot happen that lemonade will suddenly be ranked lower than orange juice by society.

The third requirement that Arrow lists is that the social welfare function not be influenced by extraneous factors. If A is preferred to B, then the sudden appearance of C should not influence one's choice between A and B. The situation can be illustrated by the following scene in a restaurant. (The anecdote is ascribed to the philosopher Sidney Morgenbesser from Columbia University.) "Today we feature apple pie and brownies," the waiter informs the customer who, after commenting on the limited choice, decides on apple pie. A few minutes later the flustered waiter returns to inform the patron that he had forgotten to mention that the restaurant also features ice cream. "In that case I will have a brownie," the customer announces after short reflection. The poor waiter is completely thrown off and rightly so. Obviously, the diner did not care one way or another about ice cream, since he did not choose it even when it was offered. But its sudden availability did reverse his choice between the two other alternatives, apple pie and brownies.

Our intuition tells us that this simply should not happen. And that was exactly the opinion that Kenneth Arrow advanced. Stipulating that a social ordering should not be influenced by unimportant options, he formulated this requirement as an axiom: the "independence of irrelevant alternatives." Neither individuals nor a group of people should, if they are rational, ever reverse their choice simply because a lower-ranked, and hence irrelevant, alternative becomes available.

As reasonable as the axiom of the independence of irrelevant alternatives sounds, it is quite a strict requirement and is not always fulfilled. In the context of voting, the axiom is often violated. Let's see how. In an election, one may prefer the ecologically minded candidate to the socially oriented candidate and would definitely vote for her if only the two of them ran for office. But when a capitalist contender suddenly decides to enter the race, one may decide to throw one's weight behind the socialist after all just so the capitalist does not come out on top. That is why Ralph Nader, presidential candidate for the Green Party, always scored so low in presidential races, even though many people supported his values. Realizing that there is no chance of his winning, perfectly rational electors vote for their next best candidate. (Not enough people were "rational," however in 2000—and I say this without any political implication. The diehard "green" voters who stayed with Nader took sufficiently many votes away from Al Gore, to hand the victory to George Bush. His record on global warming speaks for itself.) This is also why the appearance on the scene of a serious third candidate like, say, Ross Perot, can create havoc. Anyway, Arrow, like the waiter, advocates the independence of irrelevant alternatives.

The fourth condition that Arrow requires from good aggregation mechanisms is "citizens' sovereignty." This means that choices must not be imposed on the electorate. Never should a situation occur in which X is preferred by society to Y, no matter what the individuals' preferences between the two alternatives. The condition of nonimposition, as it is often called, implies that no outcome is precluded. For every possible outcome there exist individual rankings such that, when aggregated, they result in this outcome. This condition is often violated in real life. Even if all individuals prefer one alternative over another, some preferences are taboo. For example, there are taxes that are imposed on citizens, whether they like it or not, and in most states a red traffic light unequivocally means "stop," even if all drivers are of the opinion that right-hand turns entail no danger.

Finally, and most importantly, "the social welfare function must not be dictatorial." With this crucial postulate Arrow requires the aggregation mechanism to satisfy a democracy's most basic principle. On the one hand, a social welfare function is said to be dictatorial if the aggregation mechanism always parallels one specific person's preferences, no matter what the other individuals prefer. Nobody would accept that; we do not allow our choices to be dictated by anyone. On the other hand, the head of a government, once elected, calls the shots whether we like it or not. But in a democracy such a situation cannot persist forever. If the president's or prime minister's decisions do not express the will of the people—aggregated in whatever manner—he will not be reelected.

The proof starts with the definition of a set of individuals as being "decisive" for the choice between x and y, if society prefers x to y, whenever all members of the decisive set do. (This is regardless of what the other members of society prefer, and regardless of the preferences anybody has concerning the remaining alternatives.) Arrow then uses the five conditions the aggregation mechanism must fulfill to derive five consequences for decisive sets. By way of example, one of them says that "society as a whole is a decisive set." The proof is easy: if every individual in the society prefers Grappa to Amaretto, society as a whole also prefers Grappa to Amaretto. The other consequences are more difficult to explain, which is why I won't list them here. But—take Arrow's word for it—they are no more than direct consequences of the five requirements that an aggregation mechanism must fulfill.

Once the five consequences are stated, Arrow juggles around the alternatives among which choices must be made and the sets of individuals—all the time assuming that the five conditions of a rational aggregation mechanism hold. After a while he suddenly arrives at a contradiction; under certain conditions a set of individuals is simultaneously decisive and not decisive. Obviously this cannot be. The implication is that the five conditions that are required of an aggregation mechanism cannot hold simultaneously.

In a later version of his proof, Arrow replaced the monotonicity and nonimposition requirements by the somewhat weaker "Pareto condition." Its name goes back to the nineteenth-century Italian sociologist and economist Vilfredo Pareto who formulated various versions of this condition. If everybody ranks a certain alternative higher than another one, then this alternative must not be ranked lower in the social ordering. To illustrate: if everybody prefers coffee to tea, then it cannot be that society as a whole prefers tea to coffee. There is a stronger version: if every person save one is indifferent between coffee and tea, but one person prefers tea, then, for the community as a whole tea should be preferable.

The Pareto condition is one of the ingredients of democracy, as is the monotonicity requirement, in the sense that collective choice should be responsive to the preferences of the individuals. Specifically, Pareto asserted that if moving from A to B makes a single individual better off without hurting anybody else, then the community as a whole must prefer B to A. After all, the now better-off individual could compensate everybody else with his gains. The Pareto condition is closely related to the notion of Pareto efficiency, which describes an economic state in which nobody's situation can be improved without making at least one other person worse off.

Now that these five perfectly reasonable conditions have been stated, all that remains is to devise an aggregation mechanism that satisfies them. But here the efforts hit a brick wall. In chapter 5 of his booklet, Arrow gives a rigorous mathematical proof that it is not possible to devise a social welfare function whenever there are more than two alternatives from which to choose. Without fail, any attempt to aggregate the preferences of a group of people into a collective choice violates at least one of the five axioms.

The news struck like a thunderbolt. Since the times of Plato and Pliny, Llull and Kues, Borda and Condorcet, there had been hope that a mechanism to aggregate voters' preferences would eventually be discovered. Arrow's thesis put an end to such expectations. An appropriate method just cannot be found. No aggregation mechanism exists that simultaneously fulfills all five requirements.

Looking first at the bright side of things, Arrow started out by formulating a proposition that salvages whatever is salvageable. The proposition said that if there are only two options from which to choose, the method of majority decisions fulfills the requirements of an aggregation mechanism. In a possible attempt to sound upbeat in spite of the pessimistic message, Arrow called the proposition "the possibility theorem," indicating that under these very restrictive circumstances—only two alternatives from among which to choose—majority voting is an acceptable social choice mechanism. The theorem could be seen as an affirmation of the Anglo-American two-party system. It satisfies all conditions except the first one: by limiting the choice to just two alternatives, the domain is not unrestricted.

But soon Arrow had to face up to the inevitable. In Theorem 2 he formulated the central assertion of his PhD thesis, a statement that would turn the general reliance in the democratic process topsy-turvy. It said the following: if there are at least three alternatives, any aggregation method that satisfies reasonable conditions of rationality is either imposed or dictatorial. The democratic world would never be the same again. Only dictators could breathe a sigh of relief, no problem with their style of government.

Arrow required just eight pages to rigorously prove the theorem that would put a question mark to the theories of social choice, welfare economics, and political science. It should properly have been named "the impossibility theorem" and subsequently the literature usually referred to it thus, although Arrow stuck to his original term also for Theorem 2.

Of course, the majority vote of which we have grown so fond over the centuries is just one of the unacceptable aggregation mechanisms. Actually, this was nothing new. Condorcet was already aware that perfectly reasonable preference schedules among three voters or more may lead to a cycle. Hence plurality voting violates the condition of "unrestricted domain": only if certain combinations of individuals' preferences are excluded can cycles be avoided. Then Arrow really rubs it in: no other scheme of proportional representation, no matter how complicated, can remove the paradox of voting. Voters' sovereignty is simply incompatible with collective rationality.

There could be a way out of the dilemma, Arrow reminds us. By comparing and manipulating the utilities of the various electors, a communal preference ranking could be constructed arithmetically. But that path was excluded because utilities of different people cannot be added or compared. Something has to give; at least one of the five conditions must be dropped. Exclude Condition 1 (unrestricted domain) and the majority vote can serve as an aggregation mechanism that satisfies the other four conditions. But that would mean accepting the possibility of cycles. Drop Condition 4 (nonimposition) and be governed, Soviet-style, by predefined laws, regulations, taboos, and customs. Not a very enticing option. Or skip Condition 5 (nondictatorship). But who wants a dictator?

Two possibilities are left. For one, the monotonicity requirement could be dropped. But that would mean allowing a choice to be ranked lower by society after its ranking is raised by one or more individuals. This would be very counterintuitive indeed. (It would be reminiscent of the Alabama Paradox: raise the size of the House by one and get fewer representatives.) So we won't throw out that one. Finally, there is the Independence of Irrelevant Alternatives. It is the most controversial of the requirements, and one could envisage dropping it in order to salvage democracy. After all, some people, individually, do violate it. It would not be a very enticing prospect, however, since it implies irrational acts like ordering the wrong dessert or shifting one's support from Laurel to Hardy as soon as Goofy appears on the scene.

Others had recognized the dilemma and were looking for ways out of the quagmire. Let me give two examples. The Scottish economist Duncan Black considered restricting the citizens' preference schedules. Let us say the alternatives among which the people have to choose can be arranged along a line according to a parameter. For example, the range of political parties may be ordered from the extreme left, to the left, to the center, to the moderately conservative, to the very conservative. For such cases, Black proved that if each individual's preference schedule has a single peak—say he prefers the center party to the parties both on the right and on the left—then the majority vote fulfills all of Arrow's conditions, provided the number of electors is odd. The price of this is that Black restricted the domain from which citizens are allowed to choose, or the way in which they rank their choices. The voters thus violate Arrow's first condition of unrestricted domain. Furthermore, if a new party appears on the political scene that does not fit into the left-right scheme, like the greens or a gay lib party, or if some voter ranks both the extreme left and the center higher than the moderate left, Black's method of aggregating the preferences by majority vote would again fall victim to cycles.

Then, in the late 1960s and early 1970s, the Indian-born economist and philosopher Amartya Sen, winner of the 1998 Nobel Prize for economics, showed that there exist aggregation mechanisms that fulfill all of Arrow's requirements, except transitivity. (Recall that transitivity, implied by the axiom of "unrestricted domain," means that if, say, a committee for public buildings prefers an opera house to a football stadium and a football stadium to a skating rink, then the committee must prefer the opera house to the skating rink.) He investigated the implications of relaxing the transitivity requirement to quasi-transitivity (if the committee is *indifferent* between an opera house and a football stadium, and *indifferent* between a football stadium and a skating rink, it might still prefer the opera house to the skating rink). There were other heroic efforts to relax this or that requirement just a teeny little bit. But in the final analysis, all attempts to amend, append, or adjust the conditions required of a good aggregation mechanism are cop-outs.

Arrow's finding was extremely troubling. No democratic constitution exists that produces a coherent method of social choice; only a dictatorship can fulfill a handful of innocuous sounding conditions. We are caught between five rocks and a hard place. Either we accept cycles, or dictatorship, or imposed choices, or one of two kinds of irrational behavior, or we throw democracy out the door. We can't have it all ways.

*   *   *

In his pathbreaking book, Arrow showed that the preference schedules of a population of voters cannot be aggregated into a social preference ordering. If that were not bad enough already, it is only part of the story. Worse was yet to come. Arrow had taken the preference schedules of the voters for granted. But what if the electors' answers are not truthful? What if, realizing that their first choice has no chance of winning, they pretend to support a different alternative or candidate, thus pushing their second or third choice to the fore? When someone pointed out to Jean-Charles de Borda that his method could easily be manipulated by a group of electors who decide to deprive the front-runner from victory, he was indignant. "My scheme is only intended for honest men," the navy officer snapped. But what if, in the interest of the second best, the electors are not honest?

It is a problem with which Pliny the Younger had already wrestled in the first century AD and voters are still wrestling with it in the twenty-first. During the American election in 2000, for example, numerous supporters of Ralph Nader preferred to cast their ballots for Al Gore rather than voting for their true first choice, hoping at least to beat George Bush. It was to avoid such misrepresentation of their true feelings that electors in ancient times and in the Middle Ages were often obliged to take oaths that they would vote honestly. The problem of strategic voting was not addressed by Arrow. But in the early 1970s, two graduate students, the philosopher Allan Gibbard and the economist Mark Satterthwaite, independently decided to investigate the question. Specifically, they asked themselves how susceptible voting systems are to manipulation by electors. Can voters influence the outcome by misrepresenting their true intentions? They considered a somewhat simpler setting than the one Arrow investigated. While Arrow was concerned with a complete ranking of all candidates, or alternatives, from best to worst, Gibbard and Satterthwaite showed that problems arise even if only a single winner is sought.

In 1969, Kenneth Arrow, Amartya Sen, and the philosopher John Rawls announced a seminar series, run jointly by the departments of economics and philosophy of Harvard University, on "Decision Making in Organizations." It was a grandiose event with the foremost economists and philosophers from MIT and Harvard sitting in the audience week after week. Only two graduate students were present at the first meeting, one of them being Gibbard. When Arrow announced that they too would be expected to present papers during the seminar series, the other student, a personal acquaintance of this author, rushed to the registrar's office to drop the course. Gibbard stayed and when his time came to give a lecture he talked about his doctoral thesis, the manipulation of elections. Everyone present, including the other graduate student who continued to audit the series, was greatly impressed, and with that Gibbard's academic career was all but made. Four years later, in 1973, he published the landmark paper "Manipulation of Voting Schemes: A General Result," in *Econometrica*, one of the foremost journals in the field of economics. Below, I will tell more about what the paper is about.

Unbeknownst to Gibbard, Mark Satterthwaite, a doctoral student in the department of economics of the University of Wisconsin was working on a PhD thesis in the early 1970s that dealt with the exact same subject matter. Satterthwaite did not know of Gibbard's article. After all, it would only be published in 1973, which is when Satterthwaite's own thesis was accepted by the faculty of the University of Wisconsin. By the time an edited version of his thesis appeared in the *Journal of Economic Theory* it was already 1975. In fact, the first Satterthwaite heard of Gibbard's previously published work was when a referee who was checking his submission to the *Journal of Economic Theory* pointed out its existence. Nevertheless, the theorem is today known as the Gibbard-Satterthwaite Theorem, and rightly so, since both Gibbard and Satterthwaite had thought of it and worked out the proof simultaneously, albeit using different techniques. One of the differences is that Gibbard described the misrepresentation of one's preferences as a manipulation while Satterthwaite called it a strategy.

What is the sad news about the theorem? Gibbard and Satterthwaite proved that any democratic election method that purports to elect a winner from among at least three candidates can be manipulated. By misrepresenting his true preferences and pretending to prefer a candidate that he actually does not, a voter can influence the electoral outcome. No matter which election method is used–plurality, absolute majority, Borda count, two-by-two elimination, whatever—the Gibbard-Satterthwaite Theorem says that a liar may help get a candidate elected who would not have stood a chance if all electors had been truthful about their preferences. (It may take a whole coalition of liars, but if the results are very close, a single liar may suffice.) Hence, no election method exists that is both democratic and strategy-proof!

There is only one method that cannot be manipulated. By now it may come as no surprise that it is the dictatorship. Obviously, it makes no difference whether you vote honestly or dishonestly in a totalitarian regime since the dictator has his say in any case. And the dictator has no need to lie since his preference becomes law automatically.

Now, is it immoral to hide one's true preferences, thus falsifying the supposedly honest outcome? To live in society, one often has to make compromises. This happens with all aspects of daily life: which job to take, what house to buy, where to go on vacation. Many more couples would separate than actually do, if spouses would not stand back and settle for an agreeable, if not preferred, decision. Thus, a family may end up going neither to the boxing match, nor to the ballet, neither to the picnic, nor to the fancy restaurant. They may settle for a visit to the movie theater followed by a snack at the local diner, and nobody will be the worse for it.

Why should it be different in elections? Abandoning one's first choice to vote for the second is just such a compromise. Yet, setting the agenda in a committee meeting and being dishonest in order to push one's preferred alternative or candidate forward, is unfair. Say a committee made up of eleven members is to elect a new Director of Social Affairs by a knockout procedure. You and four allies prefer Alice over Bruce. Five other committee members prefer Bruce over Alice. And nobody wants Dofus, except for Mrs. Dofus. Now you do two things. First you set the agenda such that Bruce has to compete against Dofus in the first round. When the time comes to fill in the ballots, you and your friends lie about their true preference and vote for Dofus. Together with Mrs. Dofus's vote, Mr. Dofus wins the round six against five. The next round, between Alice and Dofus, will be a walkover for Alice. Now that was no compromise; it most certainly was a manipulation of the decision process.

The two professors whose names the theorem carries did not put any labels on the electors' activities. Nevertheless, Gibbard's use of the word "manipulation" has a negative connotation while their portrayal as "strategy" by Satterthwaite glosses over some of their disconcerting aspects. As so often, it all depends on the context.

* * *

First Arrow, then Gibbard and Satterthwaite. . . . The state of affairs was truly pessimistic and, unfortunately, there is no happy ending to this chapter. Furthermore, the outlook is bleak with more bad news to come. In the next chapter, where we return to the problem of apportionment of seats, we will see what else is impossible.

BIOGRAPHICAL APPENDIX

Kenneth Joseph Arrow

Arrow was born in New York in 1921 and spent his youth and student days in the Big Apple. The family lived in comfortable circumstances until the Great Depression wiped out most of their wealth; for the next ten years the family lived in poverty. Arrow passed the entrance exam to Townsend Harris High School in Queens, a school that was known for its high academic standard, and attended it from 1933 to 1936. The teachers, some of them with doctorates and hoping to become professors at a university, were of a very high caliber. It is thus not surprising that the school produced no less than three Nobel Prize winners—Arrow himself, his classmate Julian Schwinger (Nobel Prize in Physics 1965), and Herbert Hauptman (graduated high school 1933, Nobel Prize in Chemistry 1985). But when the time came for Arrow to go to college, his parents could barely afford the cost. Fortunately, City College of New York offered higher education without tuition fees, and Arrow was forever grateful for the opportunity accorded him. Even in the autobiography he would write for the Nobel Foundation more than thirty years later, he did not forget to mention "that excellent free institution." At City College he majored in mathematics with minors in history, economics, and education, and had the intention of becoming a math teacher. But when he graduated, winning the Gold Pell medal for the highest grades, there were no positions available in the New York City school system. So he entered Columbia University to continue studying mathematics. Arrow received his master's degree in 1941 and then was not sure of what he wanted to do next.

To his great fortune, he had taken a course at Columbia in mathematical economics with Harold Hotelling, a statistician who held an appointment at the department of economics. This experience proved to be propitious: Arrow decided that mathematical economics was the subject to which he would henceforth devote his life. A fellowship to the department of economics ensued but then life was interrupted by the Second World War. He entered the U.S. Air Force in 1942 as a weather officer, rising to the rank of captain in the Long Range Forecasting Group. One day, making use of their academic training, Arrow and his colleagues decided to submit their work to a statistical test. They investigated whether the group's aim—forecasting the number of rainy days one month in advance—was being attained. Not surprisingly, the conclusion was that it was not. They sent a letter to the General of the Air Force, advising the dissolution of the Long Range Forecasting Group. The response came half a year later: "The general is well aware that your forecasts are no good. However, they are required for planning purposes." So the group continued to prognosticate sunny days and rainy days using techniques that were about as good as drawing lots from a hat. Arrow left the Air Force in 1946. Something positive did nevertheless originate from Arrow's work in the Air Force; his first scientific paper, "On the Optimal Use of Winds for Flight Planning," was published in the *Journal of Meteorology* in 1949.

After the war, Arrow continued graduate work at Columbia. Mindful of the hardships that his family had suffered during the Depression, he was on the lookout for a solid, down-to-earth profession. For a while, he toyed with the idea of becoming a life insurance actuary and actually passed a series of actuarial exams. While actively searching for a job in the insurance industry an older colleague dissuaded him, and Arrow decided to embark on a career in research. In 1947, he joined the Cowles Foundation for Research in Economics at the University of Chicago. There he encountered "a brilliant intellectual atmosphere ... with eager young econometricians and mathematically inclined economists." It was also there that he met Selma Schweitzer, a young graduate student, whom he subsequently married. She was at the Cowles Foundation on a fellowship designed for women pursuing quantitative work in the social sciences. Originally the fellowship indicated preference for "women of the Episcopal Church," but the religious affiliation was subsequently dropped, which was fortunate because Selma, like Ken, was Jewish.

After earning his PhD degree Arrow was hired by the economics and statistics department at Stanford University in 1949 and then promoted through the ranks, eventually becoming Professor of Economics and Professor of Operations Research. Except for an eleven-year interlude at Harvard University, and visits to Cambridge, Oxford, Siena, and Vienna, Arrow spent his whole career at Stanford until retirement in 1991. He won numerous prizes, among them the John Bates Clark Medal in 1957, awarded every year to an outstanding economist under forty, and of course the Nobel Prize in 1972. He was elected to the National Academy of Sciences and to the American Philosophical Society and has received more than twenty honorary degrees. Even the Vatican honored him by making him a member of the Pontifical Academy of Social Sciences. During his rich career, Arrow also served on the staff of the United States Council of Economic Advisors, as president of the Econometric Society, and as fellow and member of numerous learned societies. He did not slow down after his retirement and has remained very active as an emeritus professor. During many summers, for example, he leads an advanced workshop in economic theory at the Hebrew University in Jerusalem.

In 1986 the Institute of Management Science and the Operations Research Society of America awarded Arrow the John von Neumann Theory Prize. The laudation stated that "his lightning-quick mind, his awesome wealth of knowledge, his extraordinary breadth of interests, his elegant prose and language, and his great personal warmth have inspired and charmed countless students, colleagues, and associates."

Allan Gibbard

Born in Providence, Rhode Island in 1942, Gibbard grew up in West Virginia and studied mathematics at Swarthmore College. There he received his BA degree, with minors in physics and philosophy, and then joined the Peace Corps in Africa. For two years he taught math and physics at the Achimota Secondary School, an elite high school in Accra, Ghana. Back in the United States, he returned to his college minor, philosophy, and obtained his PhD from Harvard University in 1971. As a professor of philosophy, first at the University of Chicago, then at Pittsburgh and Michigan, Gibbard tried to characterize the nature of moral judgment and define the meaning of moral statements. Gibbard has contributed important advancements to ethics, metaphysics, philosophy of language, and the theory of identity.

Philosophers are wont to rise to lofty heights of discourse, often asking questions about questions rather than answering them, thus leaving the realm of nitty-gritty, real-time decisions far behind. Gibbard fits this mold but only to a degree. Even though his books, *Wise Choices, Apt Feelings* (1990) and *Thinking How to Live* (2003), seem to promise practical advice of the how-to variety, they are nothing of the sort. And with papers like "Norms for Guilt and Moral Concepts," "Preference and Preferability," "Truth and Correct Belief," one would not expect Gibbard to stoop to down-to-earth subjects like how elections can be manipulated. But this is exactly what he did and, in the process, the mathematics he had learned as an undergraduate stood him in good stead.

Mark Satterthwaite

Satterthwaite received his BA in economics at the California Institute of Technology and then went to Wisconsin for his MA and PhD. Submitted and accepted in 1973, his thesis carried the title "The Existence of Strategy-Proof Voting Procedures." After earning his doctoral degree, he joined the faculty at the Kellogg School of Management at Northwestern University, never to leave it again, except for a semester as visiting professor at Caltech. He worked his way through the ranks, starting as an assistant professor even before he was formally awarded his PhD, all the way up to department head and to a named chair in Hospital and Health Care Management. His interests were, and still are, microeconomic theory, the economics of industrial organization, and health economics.

MATHEMATICAL APPENDIX

The Axiom of Choice

In mathematical logic there exists a very famous—and controversial—postulate called the axiom of choice, which has a subtle connection to Arrow's first axiom. Formulated in 1904 by the German mathematician Ernst Zermelo, the axiom of choice says that if one is presented with an infinite collection of sets, each of which contains a number of elements, one is always able to pick a representative member of each set. If presented with infinitely many pairs of gloves, there could be a rule that would say, "from each pair, pick the left glove." But when presented with pairs of socks, for example, there is a problem since left and right socks are indistinguishable. Thus there exists no picking rule and one finds oneself in the position of Buridan's donkey. The axiom of choice is a way out of this predicament. Without specifying a procedure or a choice rule, it simply postulates that one is able to pick an element from each set.

In mathematical proofs it is often assumed—sometimes without the prover or the reader actually being aware of it—that one can always pick a member from a group of objects.

For a finite number of sets, one may simply move from set to set and physically point to one of the objects at random. But for infinitely many sets one requires a rule. "Pick the tallest child from each class" or "pick the lowest-calorie drink from each bar" are acceptable rules. But "pick one match from each box" is questionable because it implicitly assumes that such a pick can always be made. Hence it requires the axiom of choice.

Now on to Arrow's first axiom. It says that when a decision maker is presented with two alternatives he can always make a comparison between them. Even when faced with pairs of socks, he is able to pick one sock as his preferred choice. Thus Arrow's axiom treats socks and gloves in the same manner, just as the axiom of choice does. However, there is a difference between Arrow's axiom and the axiom of choice. In addition to preferring one alternative to the other, Arrow allows indifference between them. Indifference is an acceptable choice and the decision maker won't die like Buridan's donkey.

Gödel and the Impossibility Theorem

The way in which the Impossibility Theorem threw democratic principles into turmoil reminds us of a similar event twenty years earlier. A young man in Vienna, Austria, had done to mathematics what Arrow did to social science and political theory. In 1931 the logician Kurt Gödel published a paper in the German journal *Monatshefte für Mathematik und*

Physik. It was titled *"Über formal unentscheidbare Sätze der Principia Mathematica und verwandter Systeme"* (On formally undecidable theorems in Principia Mathematica and related systems) and showed that statements exist that are true in a mathematical system but cannot be proven within its axioms. (Gödel's paper referred to Alfred North Whitehead and Bertrand Russell's monumental work *Principia Mathematica*.) Gödel's Incompleteness Theorem, as it was henceforth called, proved that under certain circumstances an axiomatic system cannot be both consistent and complete, thus putting to rest all attempts to set mathematics on an axiomatic basis. Thereby it did to mathematics what Arrow's Impossibility Theorem did to social choice theory. An aggregation mechanism, based on a handful of axioms, cannot fulfill a few reasonable requirements and be democratic. (To the general malaise one may add that Werner Heisenberg had done something similar to physics four years earlier, in 1927, with his Uncertainty Principle.)

There is a well-known anecdote about the eminent logician that may have some bearing on what we just discussed. Gödel left Austria during the Second World War and found a new home at the Institute for Advanced Study in Princeton. In 1948, when he decided that he would not return to his fatherland, he applied for American citizenship. His colleagues Albert Einstein and Oskar Morgenstern, naturalized Americans themselves, accompanied the otherworldly mathematician to the immigration office for the crucial interview. On the way they coached him on the American Constitution. Gödel, who had studied it the night before, spent the drive to the bureau arguing that the venerated document left the door open to a dictatorship. Knowing that this would not be the kind of argument the immigration official would like, Einstein and Morgenstern persuaded him not to insist on that point when questioned. Luckily Gödel heeded the warning and was duly awarded citizenship. It is not documented what the loophole was, that Gödel thought he had discovered. Was it maybe, just maybe, the idea that the democratic, one-man-one-vote majority election system that all Americans hold dear could lead to cycles, thence to a revolution and finally to a dictatorship?

THE QUOTARIANS

We return to the frustrating subject of apportionment. In the preceding chapter I recounted that Kenneth Arrow proved that any election method that satisfies reasonable conditions of rationality—like avoiding cycles—is either imposed or dictatorial, and that Allan Gibbard and Mark Satterthwaite showed that any democratic election method can be manipulated. This chapter will, unfortunately, be the bearer of further bad tidings: a fair and true allocation of seats in Congress is also a mathematical impossibility.

With the size of the House fixed at 435 in 1912, the Alabama Paradox no longer loomed. And after the inclusion of Alaska and Hawaii in 1959 no new states were likely to join the Union, so the New State Paradox also no longer posed a problem. But the population keeps growing; hence, the Population Paradox is here to stay. And of course, all the inequities that occur when rounding seats up or down remained.

Even though Congress did its best to plaster over the differences whenever possible, the problem never really went away. According to the 1950 census, California would have gained a seat, to the detriment of Kansas, if the Webster-Willcox method, a.k.a. W-W method, or Cornell method, had been used. In 1960, North Dakota would have lost one of its two seats to Massachusetts and ten years later, in 1970, Kentucky and Colorado would have gained one seat each, to the detriment of South Dakota and Montana, had the Cornell method been in force. From time to time challenges were raised. Following the 1980 census, Indiana, which stood to gain an eleventh seat under W-W, raised a ruckus. The House actually considered changing the apportionment method—New Mexico would have been the loser—but the proposal never got off the ground.

After the 1990 census, tempers flared especially high. This time the Huntington-Hill method, a.k.a H-H method, or Harvard method, caused the state of Montana to lose out by having to forfeit one of its two seats in the House. Montana did not like this at all and sued the government, specifically the Department of Commerce, which administers the apportionment of seats. Unfortunately, the W-W method would not have let Montana keep its second seat either, so the state had to dig further in order to build a case for its second seat. Lawyers did get lucky eventually, disinterring the Dean method (see chapter 10), which heretofore had never been used or even seriously considered. It had one advantage, however. It would have awarded the additional seat to Montana. (As a consequence, Washington's seats would have been reduced from nine to eight.)

The jurists built a case around the Dean method, the suit was heard in the District Court of Montana in 1991 and, lo and behold, the state was successful. A panel of three judges, with one judge dissenting, found that the method of equal proportions resulted in an unjustified deviation from the ideal of equal representation. But now the government was unhappy and the verdict was appealed to the Supreme Court. The justices in Washington DC unanimously reversed the judgment of the District Court. Justice John Paul Stevens delivered the court's opinion. He ended with the words

> the decision to adopt the method of equal proportions was made by Congress after decades of experience, experimentation, and debate about the substance of the constitutional requirement. Independent scholars supported both the basic decision to adopt a regular procedure to be followed after each census, and the particular decision to use the method of equal proportions. For a half-century the results of that method have been accepted by the States and the Nation. That history supports our conclusion that Congress had ample power to enact the statutory procedure in 1941 and to apply the method of equal proportions after the 1990 census.

To the great relief of the state of Washington, the Harvard method was vindicated once again.

After the 2000 census something surprising happened. Nobody complained. How come, did all states suddenly see the wisdom of H-H? Certainly not. But it just so happened that the W-W method would have given the exact same allocation as the H-H method for all fifty states. So nobody got upset, or at least no state could blame the method for its losses.

In spite of a lull in the disputes between proponents and opponents of the various methods, an uneasy feeling persisted. Even though the method of Huntington-Hill was tacitly accepted, mostly by default, nobody was quite at ease with the lack of theoretical underpinnings for the apportion-ment of congressional seats. It would be nice to justify the preference of one method over another by appealing to something more than to judicial expedience and political convenience. In this vacuous atmosphere aca-demics felt called upon to establish the foundations for a rigorous theory of apportionment. Two mathematicians, Michel L. Balinski and H. Peyton Young, answered the call.

Balinski, professor of mathematics at the City University of New York, had invited Young to interview for the position of assistant professor at the graduate school. The two hit it off immediately. Young was very im-pressed by the fact that Balinski had just finishing an interview with French television journalists when he drove up to his house. Apparently a person could combine mathematical research with influencing policy. When Young joined the Graduate Center of the City University of New York as a junior faculty member he and Balinski decided to collaborate. Even though both men had obtained their doctorates in mathematics, their training, teaching, and consulting work was nothing if not interdisciplin-ary. The broad experience stood them in good stead for what they were about to do.

One of Balinski's areas of expertise was integer programming, a branch of operations research. Developed before, during, and after World War II, operations research originated in the military where logistics, storage, scheduling, and optimization were prime considerations. But it soon acquired enormous importance in many other fields, for example in engi-neering, economics, and business management. While game theory, de-veloped at the same time, was mainly of theoretical interest, operations research was immediately applied to practical problems. Whenever some-thing needed to be maximized or minimized—optimized for short—and resources were constrained, operations research offered the tools to do so.

Optimization problems arise in everyday life all around us. Business, household, school, engineering, and traffic are just some of the areas where we constantly strive to maximize something: income, grade point average, profits, strength, speed, pleasure, and so on. On other occasions one may want to minimize variables like expenditures, effort, the distance one has to walk, and so forth. One of the techniques from operations re-search's toolbox is linear programming. Whenever one wants to optimize something subject to constraints—like maximizing investments subject to a budget constraint—linear programming is the tool to use.

The solution to a linear programming problem consists of real numbers—for example, the amounts of different compounds to be used in order to create an alloy of maximal strength. Unfortunately, far more intricate problems arise when the variables must be integers. Think of the fiendishly difficult Diophantine equations, which allow only integer solutions. It is very easy to give as many solutions as you like to the equation $x^3 + y^3 = z^3$, as long as the variables can be any real numbers. But if x, y, and z must be nonzero integers, solving the equation is impossible, as Fermat tried, and Andrew Wiles managed, to prove.

Something similar occurs when decision variables must be integers. For example, when an airline decides how many aircraft to buy, how many legs to fly, or how many crews to schedule, the solutions cannot be just any real numbers, they must be integers. After all, the airline cannot buy 17.6 aircraft, fly from Memphis to Dallas 2.4 times a day, and have 0.8 pilots on standby. While it is no longer difficult to find solutions to linear programming problems—George Dantzig solved that task with his Simplex algorithm in 1947—the assignment becomes vastly more complicated when solutions must be whole numbers. Tools to handle decision problems that allow only whole-number solutions do exist, however, and Balinski became one of the world's foremost experts in integer programming. This experience, together with his interdisciplinary worldview came in handy when studying the allocation of seats in Congress.

When Balinski and Young met, Balinski was in a sad mood. He had been working intensively on transforming his work on integer programming into a textbook when, one day, a fire destroyed his office and consumed all his notes and books. One can well imagine the pain and frustration that Balinski felt when he realized his labors had come to naught. The idea of resurrecting all that he had already done was extremely distasteful. Balinski began to look for a different challenge.

So when the chairman of the math department at the Graduate Center of CUNY asked for a volunteer to teach a one-semester course to two hundred freshmen, Balinski volunteered. It was an experiment the Graduate Center was going to do with undergraduate students. These were students who were not majoring in science or mathematics, and the class was likely to be the only math course they would ever attend. It was to give them a taste of the importance of mathematics. Nobody on the faculty was very keen to take on the task, but for Balinski, still smarting from the loss of his material on integer programming, this was just what he needed.

The course would be an opportunity to convey the practical importance of mathematics without the burden of having to teach a specific body of techniques. What he now required was a problem that would be accepted as significant by the students. After reflecting for a while, Balinski hit upon the perfect subject: congressional appointments. It was ideal for two reasons. First, a constitutional problem would immediately be recognized as important by the students. Second, "almost everybody is prepared to suggest a solution to the problem," Balinski recalled many years later. "It often turns out to be bad, sparking debate and confrontation, which are ideal for the classroom."

Balinski began preparing for the course. While reviewing what had been done and written on the subject, it quickly became apparent to him that the H-H method, which was still being used, was suspect, despite the solemn endorsement by the National Academy of Sciences. True to his profession as a mathematician, he decided that the axiomatic approach should be used to sort out what method was in fact the most equitable. He recruited Peyton Young to work with him on the subject.

Their collaboration eventually resulted in the widely acclaimed book *Fair Representation: Meeting the Ideal of One Man, One Vote*. Published by Yale University Press in 1982, it was the first serious scientific study of apportionment since the time when the Founding Fathers signed the Declaration of Independence—and that includes the two reports to the National Academy of Sciences. It is worth noting that the monograph was reprinted as a second edition in 2001 with identical pagination. Not many scientific texts get reprinted after twenty years, maintaining the original material except for the correction of typos. By the way, the monograph may be the only mathematics book that ever got a multipage review in a U.S. Supreme Court decision.

The stated aim of the authors was to apply mathematical reasoning to a question of public policy "similar to the axiomatic approach used in mathematics, where the object is to discover the logical consequences of certain general principles." At first blush, the book seems easy enough to read, with many historical excursions and numerical examples, but its

seeming simplicity belies its seriousness. Even though no more mathematics is required to follow the authors' reasoning than simple arithmetic,
the problems are nothing if not challenging, and the arguments nothing if
not sophisticated. But if you expect the book to give an answer to the
question, which apportioning method is best, be forewarned: there is
none.

As Kenneth Arrow had done in his work, Balinski and Young started
their search for a good method by stating the requirements it should fulfill. The first requirement is proportionality: a state with three times the
population of another state should have three times as many representatives. By the same token, if one state grows faster than another, this
should be reflected in an increase in representation. If the requirement of
proportionality is violated, we are faced with the dreaded Population Paradox. Since the Alabama Paradox and the New State Paradox no longer
threaten, this is the one remaining obstacle.

Some allocation methods violate the requirement of proportionality. In
chapter 9 we saw how Hamilton's method of allocating seats according to
the greatest fractional remainder can give a paradoxical result. Although
Virginia's population grew more than Maine's, both in absolute numbers
as in percent, it was Maine that gained a seat at Virginia's expense.

In short, fractional remainders—varying between 0 and 1—do not reflect the states' relative sizes. Hence they are an inappropriate device to
determine which states should receive additional seats. Any method that
is based in some way or another on fractional remainders suffers from the
defect of the Population Paradox. So Hamilton's method is out.

If remainder methods are not appropriate, what method is? Let's pull
the rabbit out of the hat and explain later: it's any one of the divisor methods. Recall that for this method an appropriate number is chosen—the
divisor—such that, when dividing each state's population by the divisor
and then rounding up or down, the appropriate number of seats is allocated. If the number of seats turns out to be too large or too small, a larger
or a smaller divisor is chosen and the process is repeated. The number
that results when a state's population is divided by the divisor is called the
state's quota.

As we saw in chapter 9, divisor methods abound. The only differences
between them are the cutoff points for rounding up or down. Any set of
cutoff points works! For the five traditional methods they are: Adams

method always rounds up; Jefferson's method always rounds down; Webster's rounds at the midpoint (for example, at 1.5, 2.5); Hill's at the geometric mean (for example, $\sqrt{[1 \times 2]} = 1.414$, $\sqrt{[2 \times 3]} = 2.449$); and Dean's at the harmonic mean, which is defined as the product of two numbers, divided by their average (for example, $1 \times 2/0.5[1 + 2] = 1.333$, $2 \times 3/0.5[2 + 3] = 2.4$). Balinski and Young called the fraction where the rounding takes place a signpost. Each method has its own signpost sequence. As soon as the quota passes a method's next signpost, the quota is rounded up and the state gains a seat.

TABLE 12.1
Signposts

	2 seats	3 seats	4 seats	5 seats	6 seats	Rounding scheme
			at or beyond:			
Adams	1.000	2.000	3.000	4.000	5.000	always up
Dean	1.333	2.400	3.429	4.444	5.454	at harmonic mean
Hill	1.414	2.449	3.464	4.472	5.477	at geometric mean
Webster	1.500	2.500	3.500	4.500	5.500	at arithmetic mean
Jefferson	2.000	3.000	4.000	5.000	6.000	always down

Quota rounded to (spanning header)

As an example, consider the four imaginary states IO, HJ, MU, and NK. Dividing the populations of the four states by the divisor 50,000, we get the following apportionments after rounding:

TABLE 12.2
Rounding schemes

	Population	Quota	Adams	Dean	Hill	Webster	Jefferson
IO	361,250	7.225	8	7	7	7	7
HJ	222,750	4.455	5	5	4	4	4
MU	324,100	6.482	7	7	7	6	6
NK	836,250	16.725	17	17	17	17	16

Why do these methods avoid the Population Paradox? Say a state grows, thereby passing a signpost and gaining a seat. Say another state grows more quickly. Obviously, it moves forward toward the next seat. It may not grow sufficiently to pass its own next signpost, thus not gaining an additional seat, but it could never move back past the previous signpost. This is in stark contrast to remainder methods, where states move both forward and backward beyond the point where rounding takes place. Divisor methods would never cause a faster-growing state to lose a seat, if a slower-growing state gains one. Conclusion: no Population Paradox. End of story.

With this argument, the two mathematicians showed that divisor methods of whatever variant avoid the Population Paradox. Somewhat surprisingly it turns out that divisor methods—all of them—also avoid both the Alabama Paradox and the New State Paradox. That this is the case can be shown by the signpost argument again. Balinski and Young had been seeking a way to avoid the Population Paradox. They got the annulment of the two other paradoxes as an extra benefit.

But they did more than just show that divisor methods are good at steering clear of paradoxes. They proved rigorously that divisor methods are the *only* techniques that avoid the Population Paradox. Any method of apportionment that is not based on divisors will, unfailingly, fall victim to the Population Paradox. I will not reproduce the proof here. Suffice it to say—as the authors did when they consigned the proof to the appendix—"that is the part played by mathematics."

Once it has been established that divisor methods are the only apportionment methods that should be considered, the question arises what the differences are, if any, among the rounding schemes of Adams, Dean, Hill, Webster, and Jefferson. In terms of shunning paradoxes, they are equally suitable. But we demand more from a good method than just an avoidance of ridiculous outcomes. An appropriate technique for the allocation of seats in Congress should fulfill additional requirements. The next requirement of a good allocation method is that it not be biased toward certain states.

What is meant by *bias* is a systematic tendency to favor either large states or small states. The stress is on the word *systematic*, because in any given year it is unavoidable that some states get a little more representation than they would be due, and others get a little less. As we saw in chapter 9, Jefferson's method was abandoned precisely because it consistently favored the large states. A method is considered unbiased "if the class of large states has the same chance of being favored as the class of small states," Young and Balinski declared. The true test of an unbiased method is that over the long run the advantages and disadvantages average out.

So which divisor methods—Adams, Dean, Hill, Webster, Jefferson—are unbiased, in addition to being immune to all the paradoxes? Balinski and Young attacked this question from two perspectives: historical and theoretical. They devised an index to account for the cumulative bias between 1790 and 2000 and found that Adams had a bias-index of about fifteen toward small states, Jefferson had a bias-index of about fifteen toward large states, and Dean and Hill favored small states to the lesser rate of about three and five. But the hands-down winner on the historical test was Webster's method with a cumulative bias-index of only about 0.5 toward small states.

This is not surprising. With cutoff points at 1.5, 2.5, 3.5, . . . Webster's method gives each state a 50 percent chance of being rounded up or down every time. Obviously, this averages out in the long run. In contrast, Adams's, Dean's, and Hill's cutoffs are always below 0.5—see the signposts in the table above—which favors small states for a few reasons. First, rounding up from 2.4 to 3.0 is "worth" more than rounding up from 32.4 to 33.0. Second, the cutoff points become larger, increasing toward 0.5, the larger the state. Finally, as was explained in chapter 9, rounding up necessitates an increase in the numerator, which, in turn, means that states are penalized for each seat they already have. This, in turn, implies that larger states are penalized more. Finally, Jefferson's method rounds only down, which hurts the small states for the converse of all the reasons listed above.

From among all methods that use divisors, the Webster method (which became the Webster-Willcox method after the Cornell professor Walter F. Willcox got involved) is the only one that is practically unbiased. This fact is borne out both from an empirical-historical as well as from a theoretical perspective. Balinski and Young could not hide their surprise that nobody had noticed this until then, and that Hill's method (which became the Huntington-Hill method after the Harvard professor Edward V. Huntington got involved) had been officially sanctioned by all relevant institutions. "It seems amazing therefore that Hill's method could have been chosen in 1941 . . . and that Webster's method was discarded. A peculiar combination of professional rivalry, scientific error, and political accident seems to have decided the issue." Webster had the correct insight but Willcox lacked the mathematical wherewithal to prove the point. Recall that Willcox was a statistician, a member of a social sciences department, and as such not considered a serious interlocutor by Huntington who was a mathematician and behaved as such.

Do you remember that the National Academy of Sciences had supported Huntington-Hill's method because it lay in the middle with respect to favoring small and large states? Not a very sound argument, Balinski and Young thought when summing up their thinking on the matter: "In the end, Huntington's claim bolstered by the muddle over the 'middle,' provided the scientific excuse, and straight party-line interests provided the votes." One might add that the Supreme Court also had the wool pulled over its eyes. Even though Balinski and Young's book is extensively cited in the judges' opinion in the Montana versus the Department of Commerce case, there is no mention of its conclusion.

We demanded that a suitable apportionment method be unbiased and immune to paradoxes. Webster's method fulfills both requirements and we could lean back and relax. But there is another item on Balinski and Young's wish list. A state should receive no more and no less than its fair share of seats. This sounds pretty basic, but what do they mean by "fair share"? Let us start with the "raw seats" of a state, that is, the pro-rated number of seats, fractions and all, that the state would have received, were it not for the fact that there can only be an integer number of representatives. The fair share, sometimes called the quota, is defined as the number of raw seats, rounded up or down by no more than a half.

(By the way, why can the states not have fractional numbers of representatives? The first NAS report mentions this possibility (see chapter 10) and there seems to be nothing in the Constitution that prohibits this, and one could envisage a scheme whereby a state whose fair share is 15.368 representatives would send 16 congressmen to Washington. For the purposes of voting on a bill, the first fifteen would have one vote each, while the sixteenth would count for 0.368 votes. All the states' representatives would add up to exactly 435, and all problems of apportionment would disappear. The fractional congressperson's speaking time could be pro-rated and so could his/her office budget. On average, the House would have to accommodate no more than an extra twenty-five congresspersons.)

At first blush, the fair share requirement may sound superfluous; of course, one does not round by more than one-half. But in apportionment the unthinkable does sometimes happen. Remember that if the seats of all states, after appropriate rounding, do not add up to the desired total, the divisor method instructs us to use a higher or a lower divisor and repeat the allocation process. By the time all the seats do add up to the appropriate total, it could very well be that some states had their delegation rounded not to the next integer, but to the one beyond. These states will have received more or less than their fair share; they would be "out of quota." Have a look at the following table to see how this can happen.

TABLE 12.3
Fair share

36 seats are to be apportioned to four states.

State	Population	"Raw" seats*	Rounded seats	Divisor 46,000	Rounded seats
AA	70,000	1.58	2	1.52	2
BB	112,000	2.52	3	2.57	3
CC	208,000	4.68	5	4.61	5
DD	1,200,000	27.23	27	26.30	26
Tot.	1,600,000	36	37		36

*(State's population/Total population)*36 = State's population/44,444
After pro-rating the seats (i.e., by dividing the populations with the divisor 44,444), the rounded seats add to 37 instead of 36. Hence a higher divisor is used (46,000) and—after rounding—the seats add to the desired number of 36. However, State DD receives 26 seats, which is "out of quota." (The correct quota, or "fair share," would be either 27 or 28 seats.)

Which apportionment methods fulfill the fair share requirement? One that surely does is Alexander Hamilton's method of largest remainders. After rounding the raw seats down, the method assigns the remaining seats to the states with the largest fractional remainders. By definition, all states stay within their quotas, and this is the redeeming feature of Hamilton's method that I mentioned at the end of chapter 9. We know, however, that the method suffers from the Population Paradox. (It also suffers from the Alabama Paradox and the New State Paradox but we no longer care about these.) Since the Population Paradox must be avoided by all means, only divisor methods may be considered. So which of them, Young and Balinski ask—Adams, Dean, Hill, Webster, Jefferson—guarantees that the states' delegations to Congress lie within their quotas?

Their answer is short . . . and depressing: not one of them! In fact, no divisor method exists that respects the fair share requirement. They state and prove the sad fact as a theorem: if there are four or more states, and the House has at least three more seats than there are states, then "there is no method that avoids the Population Paradox and always stays within the quota."

How about that? Thirty years after Arrow's Impossibility Theorem we are again left in a lurch. We posited just three humble prerequisites for a good allocation method: it should be unbiased, avoid paradoxes, and stay within quota. Is that too much to ask for? Yes, it is. Even if we accept a minute bias as unavoidable, Balinski and Young showed that any conceivable allocation method violates one or the other of the remaining, utterly reasonable requirements. Either it is susceptible to the Population Paradox or violations of the quota requirement cannot be excluded. (Incidentally, the Balinski-Young bibliography is an interesting example of how scientists with open minds may, in the midst of their careers, revise their points of view in a fundamental manner. At first, Balinski and Young vehemently fought for the quota method. Then they came to appreciate the divisor methods. Shortly thereafter, they completely withdrew their support for the quota method and from then on rooted for Webster.)

In classic understatement Balinski and Young speak of a "disturbing discovery." It is disturbing all right. But when one digs for the reason, it does not really surprise. The quota requirement is, after all, a very stringent condition and is easily violated. Let us see why. Rounding the raw seats of a small state entails a much greater adjustment than when the same is done for a large state. The quota of a state with 1.5 raw seats spans 66 percent (33 percent when rounded up from 1.5 to 2, and another 33 percent when rounded down). A state whose raw allocation is 41.5 has a quota that spans less than 2.5 percent. Since the requirement to stay within the quota is much more stringent for a large state than for a small one, it is not compatible with the idea that the number of seats be exactly proportional to the populations. Recall that the divisor methods entail changing the divisor whenever this is necessary to bring the allocated seats in line with the available number of seats. This has precisely the effect of moving some states' delegations beyond the rounded numbers.

So we can't have it all ways, something has to give. As the following table shows, either the Population Paradox or the quota requirement must be abandoned. Balinski and Young opt for the latter. "Achieving apportionments that accurately reflect relative changes in populations seems more important than always staying within the quota," they declare. Actually, one does not give up very much by allowing quota violations since they do not occur very frequently in practice. Comparing theoretical estimates for the five common methods, the authors concluded that the Adams and Jefferson methods violate the quota requirement nearly always, whereas the Dean method does so only in 1.5 percent of the times, and Hill's H-H method in a bit less than 0.3 percent. But the hands-down winner is, once again, Webster's method: quota violations crop up in only 0.06 percent of the times. With reapportionments occurring every ten years, Webster's W-W method will produce quota violations only once every 16,000 years on average. (Actually, the H-H method would also not be doing too badly, with a quota violation occurring on average only once every 3,500 years.)

TABLE 12.4

Method	Hamilton	Adams	Dean	Hill	Webster	Jefferson
Quota Violation	No	Yes	Yes	Yes	Yes	Yes
Paradox						
Alabama	Yes	No	No	No	No	No
Population	Yes	No	No	No	No	No
New State	Yes	No	No	No	No	No

If an apportionment method fulfills the quota requirements, it produces paradoxes; if it is immune to paradoxes it violates the quota requirement. The sad conclusion of all this is that not all items on Balinski and Young's wish list can be satisfied simultaneously. In spite of this gloomy state of affairs, the news is not all bad. There is one method that does come close to the ideal: the W-W method. "The simplest and most intuitively appealing method of all is the best one . . . it avoids the paradoxes, it is unbiased, and in practice it stays within the quota."

So why is it not in use? Balinski and Young's book was published in 1982, but in spite of all criticism, Huntington-Hill has remained the method of choice. As pointed out at the beginning of the chapter, the state of Montana challenged it after the census of 1990 but had to resort to the largely disregarded method of James Dean in its lawsuit since both H-H and W-W took away one seat. And after the 2000 census nobody had a bone to pick.

The persistence of a method that is known to be deficient is a puzzling phenomenon. We wait breathlessly to see what will happen in 2011, in 2021 . . .

BIOGRAPHICAL APPENDIX

Michel L. Balinski

Balinski was born in Switzerland into a Polish family active in international affairs. His grandfather Ludwik Rajchman, a medical doctor by profession, was a prominent socialist intellectual who devoted his life and career to the service of humanity. After World War II he would be the founder of UNICEF and the spiritual father of the World Health Organization. From Switzerland the family moved to France, but as a Jew and a prominent opponent of the Nazis, Rajchman had to flee to the United States, taking young Michel with him. There the family took on U.S. citizenship and Michel received a thoroughly American education: BA in mathematics from Williams College in 1954, MA in economics from MIT two years later, and doctorate, in mathematics again, from Princeton University in 1959.

His studies were followed by a rich career as a consultant and professor of mathematics, economics, statistics, management, decision sciences, and operations research at various universities in the United States—for a while he also served on the advisory board to the mayor of the city of New York—before he returned to France in the 1980s. He became director of the Laboratoire d'Econométrie of the École Polytechnique in Paris. Balinski was the founding editor of the journal *Mathematical Programming*, is a noted authority on mathematical optimization and operations research, and served as president of the Mathematical Programming Society from 1986 to 1989.

H. Peyton Young

Young completed his BA in general studies at Harvard University in 1966 and then went on to obtain a PhD in mathematics at the University of Michigan. After he finished his PhD he was quite fed up with the ivory tower and its lack of connection with the "real" world. Thus he chose, in 1971, to work for a study commission in Washington rather than go into academia. But after a year in DC he was pretty much fed up with the real world too (it was the Watergate era) and decided to take another look at academia. This is when he hooked up with Balinski.

Subsequently, Young taught economics, public policy, decision sciences, and business at Johns Hopkins University, the University of Maryland, and the University of Chicago. He also held positions in Europe, as visiting professor at the University of Siena in Italy, professorial fellow at Nuffield College in Oxford, and deputy chairman of system and decision sciences at the International Institute for Applied Systems Analysis in Austria. Young is a senior fellow at the Brookings Institution in Washington DC, was elected president of the Game Theory Society in 2006, and appointed professor at Oxford University a year later. Young is truly interdisciplinary. His many dozens of publications deal with various subjects in applied mathematics, economics, game theory, and political theory. In his latest research, he is concerned with the evolution of norms, conventions, and other forms of social institutions.

CHAPTER THIRTEEN
THE POSTMODERNS

In this last chapter I will describe case studies of how three different countries wrestle with apportionment and elections in light of the impossibility theorems. Every representative democracy must select delegations for legislative assemblies, composed of integer numbers of parliamentarians. The delegations represent geographical regions or political parties. Some countries have come up with unique propositions; others still experiment with adequate, if not ideal, ways of allocating seats in parliament. By way of example I will mention two countries: one of the older democracies, Switzerland, founded in 1291, and one of the newer ones, Israel, created in 1948. Finally, I will describe a new proposal for the election of a president in France, which became a democracy after the revolution at the end of the eighteenth century.

* * *

Switzerland is known as one of the world's oldest democracies. The country consists of twenty-six cantons, each one of which wants, and gets, a say in the affairs of state. (In fact, when the United States sought a framework of government for its thirteen states in the late eighteenth century, it is the Swiss model that was adopted by the Founding Fathers.) Every ten years the citizens of all the cantons elect their delegates to the Federal Council. Article 149 of the Swiss Federal Constitution stipulates (a) that the Federal Council consists of 200 representatives, and (b) that the seats are to be allocated in proportion to the population of each canton.

We know, of course, that the constitution's apodictic instructions cannot be obeyed due to the impossibility of allocating fractional seats. And we also know, of course, that all attempts to allocate the fractions are fraught with problems. As far as Switzerland is concerned, the Alabama Paradox ceased to be a problem in 1963 when the number of representatives, which was allowed to vary until then, was fixed at 200. Nobody cared about the New State Paradox (more properly: New Canton Paradox) since the last time cantons had been added to the confederation—

Geneva, Neuchâtel, and Valais—was in 1815. Then, in 1979, after many years of struggle, the Jura region—which until then had been part of the Canton Berne—became an independent canton. But that was a one-time peculiarity and Switzerland is definitely not likely to add any more cantons. Hence, the New Canton Paradox is no longer relevant.

With the size of the National Council fixed at 200, the Alabama Paradox also loomed no longer, and Switzerland could in good conscience apportion the seats to the twenty-six cantons according to Hamilton's method. (Hamilton's method is known in German-speaking Europe as the Hare-Niemeyer method after the British lawyer Thomas Hare and the German mathematician Horst Niemeyer.) The Population Paradox still threatens Switzerland, albeit without having caused any problems so far. And in the spirit of "if it ain't broke, don't fix it," the issue has been put aside. For the time being, the method of rounding down and then distributing the remaining seats according to the largest fractions is still in force.

But the problems are far from over, even after the 200 seats of the council have been apportioned to the cantons. Each canton's seats must now be assigned to the individual parties. Articles 40 and 41 of the Swiss Federal Law say how this is to be done. The key to the distribution is based on Victor D'Hondt's proposal. D'Hondt (1841–1901) was a Belgian lawyer, tax expert, and professor of civil rights and tax law at Ghent University. An impassioned advocate of the rights of minorities, he championed proportional representation and devised a method that would allow minorities to have a say in the matters of state.

D'Hondt's method, suggested in 1878, assures that the greatest number of voters stand behind each seat. It works as follows: for each seat, the number of votes cast for a party is divided by the number of the seats already allotted to the party, plus one. The seat then goes to the highest "bidder." The process continues until all seats have been filled. (Table 13.1 should make this complicated-sounding procedure somewhat clearer.)

The Swiss adopted this method but it did not take them long to realize that they had reinvented the wheel. Namely, it turned out that D'Hondt's method is computationally equivalent to the method, put forth a hundred years earlier by Thomas Jefferson, to apportion delegates to the House of Representatives in the United States. (Find a divisor such that when the numbers of votes are divided by that divisor and the results are rounded down, one obtains exactly the number of seats to be allocated.) As far as

TABLE 13.1
Jefferson-D'Hondt method

....

10 seats are to be allocated.

Votes	List A 6570	List B 2370	List C 1060
1. Seat	6570*	2370	1060
2. Seat	3285*	2370	1060
3. Seat	2190	2370*	1060
4. Seat	2190*	1185	1060
5. Seat	1642*	1185	1060
6. Seat	1314*	1185	1060
7. Seat	1095	1185*	1060
8. Seat	1095*	790	1060
9. Seat	938	790	1060*
10. Seat	938*	790	530
Total	7	2	1

The number of votes of each list is divided by the number of seats that it already has been allocated, plus one. The highest bidder (denoted by *) receives the seat, until all seats have been allocated. Let us inspect the allocation of the 3rd seat: List A already has two seats, so its population is divided by 3, which is 2190. List B and C have no seats yet, so their populations are divided by 1, giving 2370 and 1060 respectively. Since 2370 is the largest of the three numbers, List B gets the seat. For the 7th seat List A's population is divided by 6 since it already has 5 seats, List B's and C's populations are divided by 2 and by 1, respectively. The "highest bidder" for the 7th seat is List B. (Using, say 900, as a divisor, and rounding down—Jefferson's method—results in the same distribution of seats.)

the Swiss were concerned, they refused to relinquish ownership of the title to either the Belgians or the Americans and decided to name the method after Eduard Hagenbach-Bischoff (1833–1910), a homegrown professor of mathematics and physics at the University of Basel. Hagenbach-Bischoff spent most of his career studying the composition of glacier ice, the velocity of viscous fluids in pipes, fluorescence, and the propagation of electricity in wires. But he was also politically minded and served for many years on the council of the Canton of Basel. There he came across D'Hondt's method and fought vehemently for its adoption. After it was finally adopted in Basel in 1905, he fought vehemently against the use of the term "Hagenbach-Bischoff method," pointing to D'Hondt's earlier work. His efforts notwithstanding, his name got indelibly stuck to the method in Switzerland.

The fact that the Jefferson-D'Hondt-Hagenbach-Bischoff method slightly favors larger parties was not considered by the Swiss to be a major flaw. After all, when the proportional representation movement first took off

the ground, its proponents in Switzerland had to fight against the old ma-
joritarian system with its "winner take all" ideology. Hence the slight bias—even though it looks irritating when compared to the Webster method—was of no concern to the Swiss in the old days. It was assumed that
its disadvantageous impact would be felt by smaller parties only when it
is applied in a cumulative fashion, such as when the parliament uses the
D'Hondt method a second time to fill various committees.

Not everybody agreed with that assessment. Very small parties, like the
Greens or other special interest groups, felt left out. In a canton with only
a few seats in the National Council, they may not obtain any representation at all. For example, on the basis of the 2000 census, ten cantons had
four seats or less. Hence, even sizeable parties, especially if their constitu-
encies are scattered all over Switzerland, may not be able to make their
voices heard in the National Council at all. Voters will feel cheated. If a
canton has, say, two seats, the ballots of approximately a third of the voters may eventually be ignored since their parties do not make it into the
council. Many citizens would be either disenfranchised or—knowing that
their votes are in danger of being lost—would vote not according to their
conscience but, by default, for one of the larger parties.

The problem also rears its head in local elections. In the Canton of Zurich there are eighteen districts—some small, some large—and many parties compete for the Cantonal Council's 180 seats. In some districts a
dozen parties may run for as few as four seats. Dissatisfaction with the
method reached the Supreme Court of Switzerland. The judges sided with
the complainants and ordered a revision of the apportionment procedure.
Thereupon the canton entrusted an official of the Department of the Interior with the preparation of a new procedure of apportionment. It was to
be fair both to the districts and to the parties, and was to guarantee that
each vote counts.

The official did what most of us would do nowadays when charged
with a similar task: he searched the Web. Surfing around, he came across
a site by the German professor of mathematics Friedrich Pukelsheim. You
may remember him as the one who put Ramon Llull's writings on the Web
(see chapter 3). On the Web site the official found exactly what he was
looking for: a paper that dealt with the problem of how to represent districts in proportion to their populations and, at the same time, represent
parties in proportion to their total vote over all districts. In fact, what he

found was the translation into German of an article that Michel Balinski had written for the magazine *Pour la Science*, the French edition of *Scientific American*. Pukelsheim had been asked to translate the article for its sister magazine in Germany, *Spektrum der Wissenschaft*. But the professor went one better; he also developed the computer software necessary to implement Balinski's method. After the article was published, the editors allowed Pukelsheim to put it on his Web site, which is where the Swiss official found it. It dealt with the exact problem the Canton of Zurich faced.

Balinski and colleagues had developed a solution for what they called the biproportional problem. It is an ingenious method that fulfills all requirements. (I only explain a simplified version of the method. In particular, I will not adjust the weights for parties of different sizes.) First, the number of representatives of each district as a whole is determined, based on the census and distributed using Webster's method, the divisor method with standard rounding. Second, the number of representatives for each party as a whole is determined by applying Webster's method to the canton's election results. (Actually, any divisor method may be used. Balinski recommends Jefferson's rounding method in order to allocate seats to the parties because big parties are favored, which is desirable to guard against small parties.) The key question now is, which districts supply which parties' representatives. In order to answer it, a more detailed analysis of the election results is needed.

We start out with a table, or matrix, with rows representing the districts and columns representing the parties. The number of votes that each party received in each of the districts is entered into the appropriate cell. This votes-matrix will serve as a basis for the computation of another matrix; let's call it the seat-matrix. It will show which districts provide party representatives, and how many. The totals of each row and of each column of the seat-matrix are given. They are, respectively, the total number of representatives of each district and of each party, as determined previously. By filling the seat-matrix we will allocate the party representatives to the various districts.

The exercise is reminiscent of a large Sudoku puzzle, albeit with a tricky twist. In Zurich the seat-matrix would be made up of 18 rows and 12 columns—18 being the number of districts, 12 the number of parties. The totals of all rows and all columns are already given, and one now seeks appropriate numbers for each cell. One requirement is that rows and columns add up to the required totals. (The grand total of all parties, which equals the grand total of all districts, is 180, the number of council members.) That's the Sudoku part. Another requirement is that the eventual allocations somehow reflect the relative strengths of the parties in the districts. That's the tricky twist part.

Pukelsheim proved that by simultaneously using Webster's divisor method both with respect to the parties and to the districts, one arrives at a seat-matrix that fulfills all requirements. Moreover, it is the *only* seat-matrix that does so. The difficulty is that divisors cannot be computed directly but must be sought iteratively. With the computer algorithm that Pukelsheim had developed this presents no problem, however. The beauty of the method is that both the districts and the parties are represented proportionally and that every vote counts. Even votes cast for small parties in small districts—which may not lead to representation in the districts themselves—find their way into the parties' pots, thus helping the parties gain representation elsewhere.

TABLE 13.2A AND B
Biproportional method

....

(A) Votes-Matrix

	Party AA	Party BB	Party CC	Total (Census)
District 1	1800	1200	1500	4500
District 2	3600	1350	2250	7200
District 3	4500	6000	1800	12300
Total (Elections)	9900	8550	5550	24000

9 seats are to be allocated. Based on the census, Webster's method with a divisor of 2850 results in 2 seats for District 1, 3 seats for District 2 and 4 seats for District 3. Based on the election, Webster's method with a divisor of 2700 apportions 4 seats to Party AA, 3 to Party BB and 2 to Party CC. (We assume all citizens who were counted in the census also cast their votes.)

(B) Seat-matrix

	Total Seats	Party AA 4	Party BB 3	PartyCC 2	District Divisor
District 1	2	1	0	1	1.01
District 2	3	1	1	1	1.10
District 3	4	2	2	0	1.30
Party divisor		2250	2400	2775	

By applying both the Party Divisor and the District Divisor to the cells of the Votes-Matrix, and rounding, one obtains the seat-matrix. (E.g., Party AA in District 3: 4500/2250/1.30 = 1.54. After rounding, this results in 2 representatives for Party AA in District 3.)

The method was first used in Zurich in February 2006 and on the whole everybody was satisfied with the results. But is the biproportional method really fair? It is possible, after all, that a party in one district receive more seats than a larger party in the same district because surplus votes were "carried over" from elsewhere. At first blush this may seem unfair, but taken overall, every party does get its fair share of delegates. Small parties profited doubly from the new method; first, their votes in small districts were not lost and, second, citizens who had been unwilling to vote for parties that had stood no chance previously were now encouraged to cast their votes for their preferred party, even if small. Any misgivings large parties may have had as a consequence were not voiced since it would have been politically incorrect to admit to such selfish motivations.

* * *

Now on to a young democracy, Israel. Founded after World War II as a home for the Jewish people, the state has never ceased to be threatened by the neighboring Arab states. But pressures have emanated also from the inside. Immigrants from all over the world gathered in Israel, orthodox Jews from Baghdad and Warsaw, secular Jews from Paris and London, highly educated and liberal immigrants from Germany, and devout but barely literate Jews from Yemen and Morocco. In addition there are Muslims, Christians, and Bedouins. More recently, hundreds of thousands of immigrants from Ethiopia and a million immigrants from the former Soviet Union have settled in the Holy Land. The whole state is a mixture of cultures, religions, languages, and customs. And everyone, of course, wants his voice to be heard and his interests to be represented in parliament.

All this makes for a very vibrant democracy. Come election time, at least two dozen parties usually compete for the 120 seats in the Israeli parliament, the Knesset. Because of the heterogeneity of the population, an intentionally low threshold of no more than 2 percent (until recently 1.5 percent) of the valid ballots was set for entry into the Knesset. For election purposes, the whole state is considered one constituency, and usually a dozen parties or more manage to pass the hurdle and enter the Knesset—some of them with just two or three representatives. Such a fractionalized parliament makes the job of governing very difficult, and hardly any government manages to finish its term. As a consequence, elections take place every two to three years.

Because many very small parties pass into the parliament many fractional seats are lost. In the mid 1970s, two members of the Knesset, Yohanan Bader from the right side of the political spectrum and Avraham Ofer from the left, decided to do something about it. After all, if a party obtains sufficient votes for two seats and gets rounded down, every fourth vote for this party is lost. Bader and Ofer thought that the party's supporters should know that in such a case they at least contributed to the success of a somewhat like-minded party. (This does not hold for parties that do not make it past the hurdle. Votes for a party that does not reach the threshold are definitely lost.)

The method Bader and Ofer proposed was to let two political parties whose platforms vary only slightly pool their surplus votes. At least one of them should have a chance of garnering another seat. So to allocate Knesset seats to the various parties, Bader and Ofer proposed the use of a uniquely Israeli version of the Jefferson-D'Hondt-Hagenbach-Bischoff method. Before any apportionment is applied, similar-minded parties would be allowed to pool their surplus votes such that one of them may be able to obtain an extra seat.

On April 4, 1975, after a seventeen-hour debate, the Knesset voted the Bader-Ofer bill into law and since then Israel boasts the Jefferson-D'Hondt-Hagenbach-Bischoff-Bader-Ofer method of allocating seats in parliament. Before each election, like-minded parties sign "surplus votes" agreements. If after the initial allocation of the integer numbers of seats to the Knesset the two parties combined have sufficient votes left over for an additional seat, the one with fewer such surplus votes contributes them to the other party, thus enabling it to receive an extra seat. By now the method has worked to the satisfaction of all concerned for more than three decades. Maybe one reason nobody ever complained is that the numerical calculations are effected by computers behind the scenes and nobody really knows who the third party is that had to give up a seat for the benefit of one of the two who combined their surplus votes.

*　　*　　*

In France meanwhile, Michel Balinski had not kept idle. Having determined with Peyton Young that the allocation problem has no solution, he let the problem of apportionment to the French parliament be, and turned to the election of a president. Could the election procedure be improved somehow to avoid cycles in spite of Arrow's Impossibility Theorem? Together with a young colleague at the École Polytechnique, Rida Laraki, he came up with a new proposal. It was to avoid all obstacles, like Condorcet's Paradox or problems surrounding Borda's scheme, as well as Arrow's Impossibility Theorem.

The two mathematicians suggested an enhanced election procedure. Electors would no longer just put a slip of paper with the name of their preferred candidate into the ballot box. Instead, they would fill out an "evaluation form" in which they would assess all candidates, by assigning them grades ranging from "very good" to "good" to "satisfactory" and so on, all the way down to "reject." The percentage of votes each candidate received in each category would be noted. Then each candidate's so-called median would be determined: beginning with "very good" the percentages are added until the candidate has obtained at least half the electors. The candidate with the best median or, if more than one candidate have the same median, the one with the higher percentage at the median, would be declared winner.

An opportunity to test the method arose during the French presidential elections in early summer 2007. In the French presidential elections, the winner must garner at least half the votes. With three or more serious candidates usually running for office, such a result is hardly achievable, and the two candidates leading after the first round head into a second round, two weeks after the first.

A dozen candidates presented themselves, the two most important being Nicolas Sarkozy, the candidate for the right, and the socialist Ségolène Royal for the left. In the first round, on April 22, Sarkozy received 31 percent of the votes and Royal got 26. François Bayrou from the centrist Union for French Democracy came in third with 19 percent of the votes. With none of the candidates having received an absolute majority, the two frontrunners, Sarkozy and Royal, headed for the second round. Bayrou, along with nine other lower-ranked candidates, was out of the game. In the runoff election on May 6, Sarkozy beat Royal with a comfortable majority of 53 percent. He thus assumed the office of *président de la république* for a five-year term. But was he really the preferred candidate of the French people?

Balinksi and Laraki asked voters at three polling stations to fill out evaluation forms after they had cast their ordinary votes. A surprising outcome emerged. Neither of the two favorites had received the best result. It was François Bayrou, the eliminated candidate, who would have won. Sixty-nine percent of the voters assessed him as being "satisfactory" or better. Only 58 percent said the same of Royal, and Sarkozy trailed far behind with only 53 percent of the voters giving him at least a satisfactory grade. Similarly, Bayrou was rejected by only 7 percent of the voters, while the rejection rate for Royal was 13 percent and a whopping 28 for Sarkozy. The far-right candidate Jean-Marie Le Pen was assessed dead last, even though he had been ranked fourth in the first round of the traditional ballot election. While he had received 10 percent of the votes in the regular ballots, a full 75 percent rejected him on Balinski and Laraki's evaluation forms.

The two mathematicians claim that their method allows a more differentiated picture since the voters' opinions of *all* candidates are taken into account via the evaluation forms. Thus it would not suffice to get the nod from half the population; candidates must strive to get the best grades from all citizens. Ostensibly their proposed method avoids all traps. Condorcet's Paradox is circumvented because the assignment of grades does not depend on the candidates' ranking. It does not fall into the Borda trap because the addition or removal of candidates does not change the evaluation accorded the initial candidates. And it steers clear of Arrow's depressing conclusion because there is no attempt to aggregate incommensurable utilities; instead, voters express, in commonly used words, the intensity of their preferences.

This raises a host of questions about the use and interpretation of expressions. Do all voters give the same meaning to words like "good" or "acceptable"? Balinski claims that in practice it is valid to postulate the existence of common language. In judging figure skaters, for example, or wines, he believes that judges do in fact have a common language of evaluation. However, the periodic scandals at Olympic Games and other sporting events would let us believe otherwise. And why should the median be the winning criterion and not, say, the "three-quarterian," which could produce a different winner?

* * *

In addition to everything mentioned in this book, more methods and techniques have been proposed for the apportionment of seats in a parliament or for the election of a president, chairman, head honcho, or *capo di tutti capi*. I will mention two of them, the single transferable vote and approval voting.

The single transferable vote was utilized for the first time in Denmark in the nineteenth century and is applied nowadays for parliamentary elections in Ireland and Malta, and for regional and local elections in Australia, Scotland, and New Zealand. In the first stage voters rank their candidates. If a candidate is ranked first by an absolute majority of electors, he wins. If not, the candidate with the fewest top ranks is eliminated from the rankings and his votes are transferred to the next candidate. Thus, all candidates move up one rank. If again nobody gains an absolute majority, the next loser is eliminated. This process continues until one candidate has gained an absolute majority and is declared winner. Criticisms of the method include that it allows for tactical voting, that a centrist candidate may be eliminated at an early stage, that some of Arrow's requirements are violated, and that the method, when used not to elect a winner but to apportion seats in a parliament, is susceptible to the Alabama Paradox.

In approval voting, every elector is allowed to indicate on his ballot sheet one or more candidates whom he considers acceptable. Thus, he puts a check mark not only next to his preferred candidate but also next to all those he considers sufficiently capable of fulfilling the job. Each check mark counts as one vote of approval and the candidate who receives most of them is declared the winner. The candidate who receives the most overall support wins.

The advantage of this method is additional flexibility. An elector has the option of voting for the preferred candidate, even if he stands little chance of winning, and for a reasonable candidate with more promising chances. Thus, the vote is not wasted. In the American elections of 2000, for example, supporters of Ralph Nader could have indicated Al Gore as an acceptable choice. As a consequence, the latter would have been elected instead of George W. Bush, which would have more closely followed the will of the people. (Truth be told, Gore did have a majority of the popular vote anyway. It was in the Electoral College that he lost his bid for presidency.) Furthermore, misrepresenting one's true preferences brings no advantage. Thus, the objectionable habit of strategic voting will not be employed in elections that are decided by approval voting.

Approval voting is or has been used by various professional societies whose members may be expected to understand its advantages, like the Mathematical Association of America, the American Mathematical Society, the American Statistical Association, the Institute of Electrical and Electronics Engineers, and the Econometric Society. Most prominently, when the candidates for the post of Secretary General of the United Nations are presented to the members of the Security Council in a straw poll, each member indicates which candidates are acceptable and which are not. (To make the procedure sound less harsh, the ballot diplomatically says "encouraged" and "discouraged" next to the candidates' names.) When the results are in, the current secretary general holds informal consultations and further straw polls are conducted until one candidate is identified, whom every member has indicated as being acceptable. This consensus candidate is then presented to the General Assembly where the proper election—which is no more than a formality—takes place. One criticism of approval voting is that the candidate who is eventually elected may simply be the one the fewest electors disapprove, that is, the one who represents the lowest common denominator.

* * *

In concluding this book, we arrive at the sad realization that the vexing mathematics of democracy does not disappear. All methods of election and all apportionment techniques have their shortcomings. Paradoxes, quirks, mysteries, conundrums, and snags that impede perfect democratic processes are here to stay. Watch out for the next election in your country and the next apportionment of your parliament.

BIOGRAPHICAL APPENDIX

Yohanan Bader

Born in Cracow, Poland in 1905, Bader was active in the Jewish Socialist Party, the Bund. When he was about twenty, he changed his political outlook, joining a more nationalistic organization, the so-called Revisionist Zionist Movement. Bader studied law and moved to East Poland in 1939, which was then under Soviet Rule. Arrested and sentenced to hard labor in north Russia in 1940, Bader left the Soviet Union after his release in 1941, joined the Free Polish Army in 1942, and arrived in Palestine at the end of the year 1943. There he joined Etzel, a right-wing underground group that used guerrilla tactics to fight for an end to the

British mandate in Palestine. Arrested by the British authorities in 1945, he spent three years in prison. After the establishment of the state of Israel in 1948, Bader became one of founders of the right-wing Herut Movement, which later joined the Likud Party and, later still, split off again. Bader was elected to the first Knesset in January 1949 and served uninterruptedly in the opposition until 1977. He died in 1994.

Avraham Ofer

Bader's political comrade-in-arms, Ofer, was situated on the other side of the Israeli political spectrum, in the Labor Party. Born in 1922 also in Poland, Ofer's family had immigrated to Palestine when he was still a child. There he joined the prestate underground army, the Hagana, and was one of the founders of Kibbutz Hamadia in the Jordan Valley. (The Hagana would later become the official Israel Defense Forces. Etzel was its bitter rival in prestate Israel.) During the War of Independence, Ofer served as a lieutenant colonel in the Israeli navy and was the first commander of the naval base in Eilat. In the early years after the establishment of the state of Israel, Ofer became a businessman. But he had been active in politics ever since 1944, when he joined the Mapai Party, which later morphed into the Israeli Labor Party. Elected to the Knesset in 1969, Ofer was named Minister of Housing by Prime Minister Yitzhak Rabin in 1974. Sadly, he became implicated in a bribery affair that shook the party and the state to its foundations. In January 1977, before anyone was indicted or anything was proven, Ofer committed suicide.

BIBLIOGRAPHY

This bibliography is not meant to be an exhaustive overview of the literature on the topics treated in the preceding chapters. Rather, the listed items are an eclectic assortment of books and articles that may help the interested reader get a taste of the subject matters and give an indication of where to start.

Arrow, Kenneth, J. 1970. *Social Choice and Individual Values*, 2nd ed. Yale University Press, New Haven, CT.

Balinski, Michel. L., and Rida Laraki. 2007. "A Theory of Measuring, Electing, and Ranking." *Proceedings of the National Academy of Sciences*, vol. 104, pp. 8720–25.

Balinski, Michel L., and H. Peyton Young. 1975. "The Quota Method of Apportionment." *American Mathematical Monthly*, vol. 82, pp. 701–30.

Balinski, Michel L., and H. Peyton Young. 1977. "Apportionment Schemes and the Quota Method." *American Mathematical Monthly*, vol. 84, pp. 450–55.

Balinski, Michel L., and H. Peyton Young. 1980. "The Webster Method of Apportionment." *Proceedings of the National Academy of Sciences*, vol. 77, no. 1, pp. 1–4.

Balinski, Michel L., and H. Peyton Young. 2001. *Fair Representation: Meeting the Ideal of One Man, One Vote*, 2nd ed. Brookings Institution Press, Washington, DC.

Bartholdi, J., III, C. A. Tovey, and M. A. Trick. 1989. "Voting Schemes for Which It Can Be Difficult to Tell Who Won the Election." *Social Choice and Welfare*, vol. 6, pp. 157–65.

Benardete, Seth. 2001. *Plato's Laws: The Discovery of Being*. University of Chicago Press, Chicago.

Bezembinder, Thom G. G. 1981. "Circularity and Consistency in Paired Comparisons." *British Journal of Mathematical and Statistical Psychology*, vol. 34, pp. 16–37.

Black, Duncan. 1958. *The Theory of Committees and Elections*. Cambridge University Press, Cambridge.

Black, Duncan, Iain McLean, Alistair McMillan, and Burt Monroe. 1996. *A Mathematical Approach to Proportional Representation: Duncan Black on Lewis Carroll*. Springer-Verlag, Heidelberg.

Bliss, G. A. 1952. "Autobiographical Notes." *American Mathematical Monthly*, vol. 59, pp. 595–606.

Bliss, G. A., E. W. Brown, Luther P. Eisenhart, and Raymond Pearl. 1929. "Report to the President of the National Academy of Sciences." National Academy of Sciences, Washington, DC.

Bonner, Anthony. 1994. *Doctor Illuminatus: A Ramon Llull Reader*. Princeton University Press, Princeton, NJ.

Borda, Jean-Charles de. 1784. *Mémoire sur les élections au scrutiny*. Histoire de l'académie royale des sciences, Paris.

Brams, Steven J. 2008. *Mathematics and Democracy*. Princeton University Press, Princeton, NJ.

Brams, Steven J. 2008.*The Presidential Election Game*. A. K. Peters, Wellesley, MA.

Caritat, Marie Jean Antoine Nicolas, Marquis de Condorcet. 1785. *Essai sur l'application de l'analyse à la probabilité des décisions rendues à la pluralité des voix*. L'imprimérie Royale, Paris.

Caritat, Marie Jean Antoine Nicolas, Marquis de Condorcet. 1793. "Sur les élections." *Journal d'instruction sociale*, Paris, pp. 25–32.

Cusa, Nicholas de, and Paul E. Sigmund (translator). 1996. *Nicholas of Cusa: The Catholic Concordance*. Cambridge University Press, Cambridge.

David, H. A. 1959. "Tournaments and Paired Comparisons." *Biometrika*, vol. 46, pp. 139–49.

David, H. A. 1988. *The Method of Paired Comparisons*. Charles Griffin, London.

Dodgson, Charles Lutwidge, and Francine Abeles. 1994. *The Political Pamphlets and Letters of Charles Lutwidge Dodgson and Related Pieces: A Mathematical Approach*. Lewis Carroll Society of North America, New York.

Durand, E. Dana. 1947. "Tribute to Walter F. Willcox." *Journal of the American Statistical Association*, vol. 42, pp. 5–10.

Farquharson, Robin. 1972. *The Theory of Voting*, Basil Blackwell, Oxford.

Felsenthal, Dan, and Moshe Machover. 1992. "After Two Centuries, Should Condorcet's Voting Procedure Be Implemented?" *Behavioral Science*, vol. 37, pp. 250–74.

Gibbard, Allan. 1973. "Manipulation of Voting Schemes: A General Result." *Econometrica*, vol. 41, pp. 587–601.

Goldenweiser, E. A. 1939. "Joseph A. Hill." *Journal of the American Statistical Association*, vol. 34.

Huckabee, David. C. 2000. "The House Apportionment Formula in Theory and Practice." *Congressional Research Service Report for Congress*, Library of Congress, Washington, DC.

Hunt, Lynn. 1996. *The French Revolution and Human Rights: A Brief Documentary History*. Bedford Books of St. Martin's Press, Boston.

Huntington, Edward V. 1921. "The Mathematical Theory of Apportionment of Representatives." *Proceedings of the National Academy of Sciences*, vol. 7.

Huntington Edward V. 1928. "The Apportionment of Representatives in Congress." *Transactions of the American Mathematical Society*, vol. 30, pp. 85–110.

Huntington, Edward V. 1928. "The Reapportionment Bill in Congress." *Science*, vol. 67, no. 1742, pp. 509–10.

Huntington, Edward V. 1928. "The Apportionment Situation in Congress." *Science*, vol. 68, no. 1772, pp. 579–82.

Huntington, Edward V. 1929. "Reply to Professor Willcox." *Science*, vol. 66, no. 1784, p. 272.

Huntington, Edward V. 1929. "The Report of the National Academy of Sciences on Reapportionment." *Science*, vol. 69, no. 1792, pp. 471–73.

Huntington, Edward V. 1941. "The Role of Mathematics in Congressional Apportionment." *Sociometry*, vol. 4, pp. 278–301.

Kelly, J. S. 1987. "An Interview with Kenneth J. Arrow." *Social Choice and Welfare*, vol. 4, pp. 43–62.

Laplace, Pierre-Simon. 1820. *Théorie analytique des probabilities*. Mme Veuve Courcier, Imprimeur-Libraire pour les Mathématiques, Paris.

Leonard, William R. 1961. "Walter F. Willcox: Statist." *American Statistician*, vol. 15, pp. 407–10.

Lines, Marji. 1986. "Approval Voting and Strategy Analysis: A Venetian Example." *Theory and Decision*, vol. 20, pp. 155–72.

McLean, Iain, and Fiona Hewitt 1994. *Condorcet: Foundations of Social Choice and Political Theory*. Edward Elgar Publishing, Aldershot, UK.

McLean, Iain, and J. London. 1990. "The Borda and Condorcet Principles: Three Medieval Applications." *Social Choice and Welfare*, vol. 7, pp. 99–108.

McLean, Iain, and J. London 1992. "Ramon Llull and the Theory of Voting." *Studia Llulliana*, vol. 32, pp. 21–37.

McLean, Iain, and Haidee Lorrey. 2001. "Voting in Medieval Universities and Religious Orders." Mimeo, UCLA Center for Governance, Los Angeles.

McLean, Iain, and Arnold B. Urken 1995. *Classics of Social Choice*. University of Michigan Press, Ann Arbor.

Morrow, Glenn R. 1960. *Plato's Cretan City: A Historical Interpretation of the Laws*. Princeton University Press, Princeton, NJ.

Morse, Marston, John von Neumann, and Luther P. Eisenhart. 1948. "Report to the President of the National Academy of Sciences." National Academy of Sciences, Washington, DC.

Pangle, Thomas L. 1988. *The Laws of Plato*. University of Chicago Press, Chicago.

Pukelsheim, Friedrich 2001. "Die Mathematik der Wahl." *Neue Zürcher Zeitung*, vol. 18/19, August, p. 51.

Pukelsheim, Friedrich. 2002. "Auf den Schultern von Riesen: Llull, Cusanus, Borda, Condorcet, et al." *Litterae Cusanae*, vol. 2, pp. 3–15.

Saari, Donald G. 1978. "Apportionment Methods and the House of Representatives." *American Mathematical Monthly*, vol. 85, pp. 792–802.

Saari, Donald G. 1987. "The Source of Some Paradoxes from Social Choice and Probability." *Journal of Economic Theory*, vol. 41, pp. 1–22.

Saari, Donald G. 1988. "Symmetry, Voting, and Social Choice." *Mathematical Intelligencer*, vol. 10, pp. 32–42.

Saari, Donald G. 1989. "A Dictionary for Voting Paradoxes." *Journal of Economic Theory*, vol. 48, pp. 443–75.

Saari, Donald G. 1991. "Calculus and Extensions of Arrow's Theorem." *Journal of Mathematical Economics*, vol. 20, pp. 271–306.

Saari, Donald G. 1994. *Geometry of Voting*. Springer-Verlag, Heidelberg.

Saari, Donald G. 2000. "Mathematical Structure of Voting Paradoxes I: Pairwise Votes." *Economic Theory*, vol. 15, 1–53.

Saari, Donald G. 2000. "Mathematical Structure of Voting Paradoxes II: Positional Voting." *Economic Theory*, vol. 15, 55–102.

Saari, Donald G., and Jill van Newenhizen. 1988. "The Problem of Indeterminacy in Approval, Multiple, and Truncated Voting Systems." *Public Choice*, vol. 59, pp. 101–20.

Satterthwaite, Mark Allen. 1975. "Strategy-Proofness and Arrow's Conditions: Existence and Correspondence Theorems for Voting Procedures and Social Welfare Functions." *Journal of Economic Theory*, vol. 10, pp. 187–210.

Saunders, Trevor J. 1972. "Notes on the Laws of Plato." Bulletin Supplement No. 28, Institute of Classical Studies, University of London, pp. 33–49.

Saunders, Trevor J. 1992. "Plato's Later Political Thought." In R. Kraut, ed., *The Cambridge Companion to Plato*, pp. 464–92. Cambridge University Press, Cambridge.

Staveley, E. S. 1972. *Greek and Roman Voting and Elections*. Cornell University Press, Ithaca, NY.

Strauss, Leo. 1975. *The Argument and the Action of Plato's Laws*. University of Chicago Press, Chicago.

Szpiro, George G. 2006. *The Secret Life of Numbers*. Joseph Henry Press, Washington DC.

Taylor, Alan D. 2002. "The Manipulability of Voting Systems." *American Mathematical Monthly*, vol. 109, pp. 321–37.

Todhunter, I. 1865. *The History of Probability from the Time of Pascal to That of Laplace*. Macmillan, Cambridge and London.

U.S. District Court. 1991. *Montana v. United States Department of Commerce* (1991), 775 F. Supp. 1358 (U.S.D.C. Mt.).

U.S. Supreme Court. 1992. *United States Department of Commerce v. Montana* (1992), 112 S. Ct. 1415.

Willcox, Walter F. 1929. "The Apportionment Situation in Congress." *Science*, vol. 69, no. 1787, pp. 163–65.

Willcox, Walter F. 1942. "An Untried Method of Federal Reapportionment." *Science*, vol. 95, no. 2472, pp. 501–3.

Young, H. Peyton. 2001. "The Mathematics of One Person, One Vote." *American Physical Society News*, vol. 10, no. 4.

INDEX

INDEX

Poundstone, William, 163
P-problems, 116
"Preference and Preferability" (Gibbard), 184
Principia Mathematica (Whitehead and Russell), 186
probability theory, 78–79, 82, 98
proportionality, 192
Pukelsheim, Friedrich, 40, 42, 205–6, 207
Pyrilampes, 1

quasi-transitivity, 178
quick sorting, 56
quota (fair share) method, 196–99, 197 (table), 199 (table)

Rajchman, Ludwik, 200
RAND Corporation, 167
Randolph, Edmund, 122
ranking, 43–44, 45, 54–58, 71–72, 82–83, 94–95
Rawls, John, 179
Reif, Wolfgang, 42
The Republic (Plato), 2–3, 6, 17–20
Republicans vs. Democrats on allocating Congressional seats, 148, 150–51, 154–55
Robespierre, Maximilien François Marie Isidore de, 76
rock, paper, scissors, 45
"The Role of Mathematics in Congressional Apportionment" (Huntington), 152–55
Roman Empire, 22, 23, 24
Roosevelt, Franklin D., 148, 155
roulette method, 157–58
rounding at the geometric mean, 137–38, 140, 164
rounding-down method. *See* divisor method
rounding-down method, two-step. *See* largest fractional remainders method
rounding-to-the-nearest number method. *See* major fractions method
rounding-up method. *See* smallest divisors method
Royal, Ségolène, 210–11
runoffs, 29, 98
Ruprecht, Theresa Anna Amalie Elise, 162
rural states' resistance to reapportionment, 147–48, 152
Russell, Bertrand, 101; *Principia Mathematica*, 186

Sankt Nikolaus Hospital library (Bernkastel-Kues, Germany), 42, 52
Sarkozy, Nicolas, 210–11
Sarret, M., 87

Satterthwaite, Mark, ix, 179, 180; "The Existence of Strategy-Proof Voting Procedures," 184
Saunders, Trevor J., 14
Schelling, Thomas, 167
Schultz, Theodore W., 168
Schwinger, Julian, 182
Science, 139, 140–41, 142, 145
Scotland, elections in, 212
Seaton, C.W., 128
secrecy vs. openness, 54–55
Secretary General of the United Nations, 213
Sen, Amartya, 178, 179
seven, significance of, 37
shell sorting, 56
showdowns. *See* knockouts
SimCity (computer program), 4
Simon, Herbert, 167, 168
Simplex algorithm, 190
single transferable vote, 212
the single transferable vote, 212
smallest divisors method, 125–27, 126 (table), 144, 192–93, 193 (table), 195
smallest divisors/range method, 155–57
Smith, Adam, 75; *Wealth of Nations*, 74
"Social Choice and Individual Values" (K. Arrow), 168, 170–79
social choice theory, 176–78
Social Psychology Quarterly (formerly *Sociometry*), 152–55
social statistics, 158
social welfare function, 172–76
Sociometry (later *Social Psychology Quarterly*), 152–55
Socrates, 1–2, 16. *See also The Republic*
soldiers, Plato on, 18–19
Sollers, 24–25
Soviet Union-United States relations, 167–68
St. Petersburg Paradox, 167
statistics, 158
Stevens, John Paul, 188
strategic voting, 68, 92–96, 108, 179
Suard, Amélie and Jean-Baptiste, 88
submarine construction, 61
Sudoku puzzles, 206–7
"Suggestions as to the Best Method of Taking Votes" (Dodgson), 110–11, 113
Supreme Court (ancient Greece), 16
"*Sur les élections*" (J.-M. de Condorcet), 81–83
surplus votes method, 209
surveying, 70–71
swaps in voting, 115